Springer Series in

MATERIALS SCIENCE

94

Springer Series in
MATERIALS SCIENCE

Editors: R. Hull R. M. Osgood, Jr. J. Parisi H. Warlimont

The Springer Series in Materials Science covers the complete spectrum of materials physics, including fundamental principles, physical properties, materials theory and design. Recognizing the increasing importance of materials science in future device technologies, the book titles in this series reflect the state-of-the-art in understanding and controlling the structure and properties of all important classes of materials.

74 **Plastic Deformation in Nanocrystalline Materials**
By M.Yu. Gutkin and I.A. Ovid'ko

75 **Wafer Bonding**
Applications and Technology
Editors: M. Alexe and U. Gösele

76 **Spirally Anisotropic Composites**
By G.E. Freger, V.N. Kestelman, and D.G. Freger

77 **Impurities Confined in Quantum Structures**
By P.O. Holtz and Q.X. Zhao

78 **Macromolecular Nanostructured Materials**
Editors: N. Ueyama and A. Harada

79 **Magnetism and Structure in Functional Materials**
Editors: A. Planes, L. Manõsa, and A. Saxena

80 **Ion Implantation and Synthesis of Materials**
By M. Nastasi and J.W. Mayer

81 **Metallopolymer Nanocomposites**
By A.D. Pomogailo and V.N. Kestelman

82 **Plastics for Corrosion Inhibition**
By V.A. Goldade, L.S. Pinchuk, A.V. Makarevich and V.N. Kestelman

83 **Spectroscopic Properties of Rare Earths in Optical Materials**
Editors: G. Liu and B. Jacquier

84 **Hartree–Fock–Slater Method for Materials Science**
The DV–X Alpha Method for Design and Characterization of Materials
Editors: H. Adachi, T. Mukoyama, and J. Kawai

85 **Lifetime Spectroscopy**
A Method of Defect Characterization in Silicon for Photovoltaic Applications
By S. Rein

86 **Wide-Gap Chalcopyrites**
Editors: S. Siebentritt and U. Rau

87 **Micro- and Nanostructured Glasses**
By D. Hülsenberg, A. Harnisch, and A. Bismarck

88 **Introduction to Wave Scattering, Localization and Mesoscopic Phenomena**
By P. Sheng

89 **Magnetoscience**
Magnetic Field Effects on Materials: Fundamentals and Applications
Editors: M. Yamaguchi and Y. Tanimoto

90 **Internal Friction in Metallic Materials**
A Reference Book
By M.S. Blanter

91 **Time-dependent Mechanical Properties of Solid Bodies**
A Theoretical Approach
By W. Gräfe

92 **Solder Joint Technology**
Materials, Properties, and Reliability
By K.N. Tu

93 **Materials for Tomorrow**
Theory, Experiments and Modelling
Editors: S. Gemming, M. Schreiber, and J-B. Suck

94 **Magnetic Nanostructures**
Editors: B. Aktaş, L. Tagirov, F. Mikailov

B. Aktaş · L. Tagirov · F. Mikailov
(Eds.)

Magnetic Nanostructures

With 130 Figures and 5 Tables

Prof. Dr. Bekir Aktaş
Gebze Institute of Technology, Physics Dept.
P.O.Box 141, 41400 Gebze-Kocaeli, Turkey
E-mail: aktas@gyte.edu.tr

Dr. Faik Mikailov
Gebze Institute of Technology, Physics Dept.
P.O.Box 141, 41400 Gebze-Kocaeli, Turkey
E-mail: faik@penta.gyte.edu.tr

Prof. Dr. Lenar Tagirov
Kazan State University
Dept. Theoretical Phsyics
Ul. Kremlevskaya 18, 420008 Kazan
Russia
E-mail: lenar.tagirov@ksu.ru

Series Editors:

Library of Congress Control Number: 2006936493

ISSN 0933-033X

ISBN 978-3-540-49334-1 Springer Berlin Heidelberg New York

Springer is a part of Springer Science+Business Media.
springer.com

Typesetting: Digital data supplied by editors
Production: LE-TEX Jelonek, Schmidt & Vöckler GbR, Heidelberg, Germany
Cover production: Manfred Bender, WMX Design GmbH, Heidelberg

Printed on acid-free paper SPIN: 11311102 57/3100/YL 5 4 3 2 1 0

Preface

The rate of development in modern digital computer systems and software has led to an almost insatiable demand for ever increasing storage capacities. In response to this demand the memory manufactures and, in particular the disk drive manufacturers have over the last decade or so come forward with spectacular increases in storage capacities and densities. The field of magnetic memory has been on the frontier of advanced materials development for many years. The momentum now gained in the technology is such that storage densities are increasing at something like one hundred percent per annum and this rate may still be rising. Over the past decades dramatic progress has been made in magnetic storage systems. In the past few years areal densities, currently above 50 Gbits/in^2, have doubled every 10 months and are expected to reach 100 Gbits/in^2 in the near future. The price per megabyte has decreased by a factor of 10 in the last decade and is predicted to drop to USD 0.05 ber megabyte within the next 10 years. The reason is that the intensive investigations in the field of the nanoscale magnetic materials promote to the great progress in various kinds of the magnetic storage media (computer floppy disks, sound/video tapes, etc.).

Among the magnetic storage devices, the hard disk drive (HDD) is the dominant secondary mass storage device for computers, and very likely also for home electronic products in the near future. The HDD is an integration of many key technologies, including head, medium, head-disk interface, servo, channel coding/decoding, and electromechanical and electromagnetic devices. Among them, the read head is the only component that has experienced the most changes, including some revolutionary ones in terms of both the operating principle and the structural design and fabrication processes during the last decade. The ever-increasing demand for higher areal densities has driven the read head evolving from a thin-film inductive head to an anisotropic magnetoresistive (AMR) head. The technical progress of last years in the preparation of multilayer thin films promote to discovering the Giant Magnetoresistance (GMR) phenomena, consisting in extraordinary changing of resistivity/impedance of the material while applying external magnetic field. The GMR- materials are already found applications as sensors of low magnetic field, computer hard disk heads, magnetoresistive RAM chips etc. The "read" heads for magnetic hard disk drives (HDD) have allowed to increased the storage density on a disk drive from 1 to 20 Gbit per square inch, merely by the incorporation of the new GMR materials. The market only in the field of GMR-nanotechnology is estimated over 100 billion dollars annually.

Magnetic recording has dominated the area of peripheral information storage ever since the beginning of the computer era a half century ago, with tapes and disks representing the two main application areas. The embodiment of the technology involves a relatively thin magnetic layer supported by a flexible or rigid substrate, which can be magnetized by an external magnetic field and which retains its magnetization after the field is removed. Information is recorded in the form of oppositely magnetized regions in the surface layer of the medium utilizing the fringing field of an inductive transducer. Reading is done by the same or a similar inductive transducer or by a magnetoresistive sensor. The magnetic recording media used in information storage can be divided into two classes. One of them is a particulate media usually employed in tapes and flexible disks. They consist of distinct physical particles of magnetic oxides or metals in an organic binder coated on to flexible polymer substrates. The other class thin film media of magnetic metals or oxides deposited by sputtering or thermal evaporation on to rigid disk substrates or flexible polymer substrates. Rigid disk systems exclusively employ sputtered metallic thin film media. They consist of small magnetic grains separated from each other by some non-magnetic phase in order to reduce intergrain exchange coupling which represents one of the main contributors to media noise in thin film media.

This book is intended to provide an updated review of nanometer-scale magnetizm for data storage applications, with emphasis on the research and application of nanoscaled magnetic materials in ultra-high-density data storage. The idea for this book was born at the NATO Advanced Research Workshop on Nanostructured Magnetic Materials and their Application (NATO ARW NMMA2003), held in Istanbul (Turkey) on July 1–4, 2003. The contributions are concentrated on magnetic properties of nanoscale magnetic materials for ultra-high-density data storage, especially on fabrication, characterization and the physics behind the behavior of these structures.

We would like to thank all the authors for their contributions. We should also acknowledge the great efforts of NATO Scientific and Environmental Affairs Division, and Turkish Scientific and Technical Research Council (TUBITAK), Gebze Intitute of Technology with its president – Prof. Alinur Buyukaksoy, and others who made major contributions to the organization of the meeting and made this publication possible.

Gebze Institute of Technology,
October 2006

Bekir Aktaş
Lenar Tagirov
Faik Mikailov

Contents

Part I U-H-D Recording Media

Part II Nano-Patterned Media

Part III Functional Elements of MRAMs

Part IV GMR Read Heads and Related

Part I

U-H-D Recording Media

Materials Challenges for Tb/in^2 Magnetic Recording

Ernesto E. Marinero

Hitachi San Jose Research Center, 650 Harry Road, San Jose, CA 95120, USA

Summary. Magnetic recording technology aims to provide the capability of storing as much as 10^{12} bits/in^2 (1.6×10^{11} bits/cm^2) in the foreseeable future. This remarkable storage density projection is made possible by recent major improvements in the microstructural and intrinsic magnetic properties of the thin film materials utilized for both recording and reading nanoscale magnetic domains which form the basis for digitally encoding information. To meet the Tb/in^2 goal, further reductions in the recording medium grain size, the grain size distribution as well as increments in the magnitude of the magnetic anisotropy will be required. Currently used longitudinal recording materials are unlikely to support such high density targets and a migration to perpendicular recording is expected. On account of the superparamagnetic limit and the need for stringent control on the domain size, it is also likely that a transition to patterned media will also be required. The read sensor sensitivity will also need to increment to provide large enough signals needed for low error rate. Therefore, breakthroughs in magnetic thin films for the disk and head components will be required to meet the technology goals.

1 Introduction

The key elements of a hard disk drive are: the recording medium, the write and read head elements, the mechanical actuator, the signal processing devices and the ancillary electronic components to record, read and to perform reliable seeking operations. Magnetic thin films play a key role in enabling this technology. They constitute the storage medium and are the key elements of the write and read head elements. Data is recorded by locally altering the direction of the magnetization in the recording medium via the magnetic field generated when current is passed through the lithographically defined coils within the recording head structure. The bit size is determined by the geometry of such write head. Data is read and processed by sensing the out-of-plane magnetic flux arising at the boundaries between recorded bits. These small magnetic fluxes are detected by utilizing thin film read heads that exploit the magneto-resistive effect when subjected to a magnetic excitation. Today's read heads which are mainly spin valves employ sensor materials exhibiting giant magnetoresistance.

To increase the storage density, the magnetic bit size and the inter-bit separation along and across data tracks must be reduced. This scaling approach has worked out very well with only minor changes in the structural and magnetic properties of the recording layers up to storage densities around 1 Gb/in^2 . Significant improvements in reducing the grain size and the intergranular exchange coupling have permitted the extension of longitudinal recording alloys up to storage densities around 100 Gb/in^2. To circumvent losses in flux amplitude that are inherent to the 1000-fold reduction of the bit size, higher sensitivity sensor materials have been developed and the physical spacing between the topmost layers of the recording of the medium and the read sensor has been reduced from $\sim$ 100 nm to $\sim$ 10 nm. Flying reliably at such small distances imposes stringent demands on the thickness and smoothness of the recording thin film structures. In particular, the overcoat (typically amorphous carbon) thickness must be reduced to a minimum ($<$ 5 nm) while preserving its mechanical integrity and strength to provide not only environmental protection to the recording layers, but also mechanical robustness. It is obvious from these overall considerations on magnetic recording technology, that the microstructural control at the nanoscale level of the multiplicity of layers that comprise the recording medium as well as the elements of the read and write head, will play a critical role in extending the technology to and beyond Tb/in^2.

The workshop addressed many facets on fundamental magnetism and materials growth and characterization that are critical to sustaining the viability of magnetic recording technology into the XXIst century.

2 Materials Challenges

The microstructural and magnetic properties of the thin films employed in magnetic storage technology will determine to a large extent the attainable density recording limits for magnetic recording. Consider first, magnetic thin films for storing the magnetic information. Increasing linear density requires that the domain wall between recorded transitions be decreased. Recording at 1.0 Tb/in^2 will require bit dimensions of $\sim$ 15 nm in diameter (for cylindrical bits) and a bit spacing of $\sim$ 10 nm. It is highly unlikely that current granular media which consists of magnetic grains segregated by a secondary non-magnetic phase will meet these requirements. The smallest magnetic grain diameter attainable with present sputter deposition fabrication methods is of the order of 5 to 7 nm. In addition, the grain size distribution of said recording materials is inadequate for meeting the stringent requirements for recording at the Tb/in^2 regime. Therefore a migration to fabrication methods or growth techniques that provide a regular array of magnetic nanoparticles is of paramount importance. Utilizing lithographic techniques and current thin film growth technologies is a viable approach. Nevertheless, it should be recognized that the minimum feature sizes needed, as well as the spatial distribution tolerances for Tb/in^2 are so stringent that current lithographic techniques will not satisfy said requirements and tolerance limits. Breakthroughs in lithography will be needed to meet the aforementioned density goals. Self-assembly of nanoparticle arrays provides another approach for generating the desired characteristics of the recording medium. Significant improvements in this area are discussed in these proceedings. Some of the challenges for self-assembly growth methods to be of practical use for magnetic recording technology pertain to the need to provide large areas of self-assembled particles exhibiting the spatial distribution that have been achieved over small areas, as well as the need to develop long range magnetic order of the nanoparticle assembly.

It is important to realize that the island size and thus, the magnetic volume cannot be arbitrarily reduced. This is on account of the superparamagnetic effect. When a single isolated magnetic particle reaches a minimum critical volume, the magnetocrystalline anisotropy is no longer sufficient to overcome the thermal energy that tends to randomize the magnetic axis orientation. This limit indicates that isolated spherical Co particle of 7.6 nm in diameter become superparamagnetic at room temperature. The limit for a given ferromagnetic material, is determined by the magnitude of its magnetic anisotropy energy. Therefore, future recording thin films will need to employ higher anisotropy materials than current Co-based alloys. A fertile area of investigation is the permanent magnet field in which anisotropies exceeding 10 to 200 times that of current recording materials have been reported. Similarly, ordered alloys such as FePt could in principle provide stable islands down to $\sim$ 3 nm in diameter. A critical caveat in the practical utilization of said high anisotropy materials is their writability. Because of their high intrinsic anisotropy, the writing coercivity of said materials is too high to permit current write head devices to control their magnetization direction. It will become either necessary to employ some form of thermomagnetic recording to reverse the magnetization direction. Applying a heat pulse during the write operation, permits a drastic reduction in coercivity, thereby allowing current write head devices to be used concurrently with the heat pulse. Alternatively, nanoscale spring magnet materials could in principle be employed. Said materials comprise two ferromagnetic structures which are strongly exchanged coupled. One component is magnetically soft and the other is magnetically hard. The soft material easily responds to the external field and through a magnetic torque effect is able to drag the magnetization of the hard magnet component.

In the area of the read sensor, significant advances will also be required for Tb/in^2 applications. The GMR response needs to be driven to higher values to compensate for losses in magnetic flux as the domain size is reduced. Novel spin-valve geometries including CIP need to be explored to extend GMR coefficients. Additional impetus for the development of higher sensitivity sensor materials has been injected by recent developments in magnetic tunnel junction devices and other spintronic materials and devices. However, as the feature sizes in read sensor materials are continuously decreased, thermal fluctuations of the magnetic order in said magnetic materials will play a key role in limiting their extendibility to ultra-high density recording. Said fluctuations labeled as magnetic-noise could become dominant in some magnetic read sensor devices with further scaling reductions.

Thus, the future evolution of magnetic storage technology critically depends on major advances in thin films physics, materials development and fabrication engineering. The proceedings of this workshop provide a wide range of contributions that seek to provide fundamental understanding and address solutions to such critical materials challenges.

Scanning Hall Probe Microscopy: Quantitative & Non-Invasive Imaging and Magnetometry of Magnetic Materials at 50 nm Scale

Ahmet Oral

Bilkent University, Department of Physics, 06800 Ankara, Turkey

Summary. Scanning Hall probe microscopes have proven themselves to be quantitative and non-invasive tools for investigating magnetic samples down to 50 nm scale. They can be run in a wide range of temperatures and have unprecedented field resolution which is not affected by external fields. Local magnetometry of very small volumes is also feasible opening up very promising avenues for characterization of magnetic nanostructures.

1 Introduction

Scanning Hall Probe Microscopy (SHPM) [1, 2] is a quantitative and non-invasive technique for imaging localized surface magnetic field distribution on sample surfaces with high spatial and magnetic field resolution of $\sim$ 50 nm & 70 mG/$\sqrt{\text{Hz}}$, over a wide range of temperatures, 30 mK–300 K. This new technique offers great advantages and complements the other magnetic imaging methods like Magnetic Force Microscopy (MFM) [3], Magnetic Near Field Scanning Optical Microscopy [4] and Kerr Microscopy [5]. In SHPM, a nano-Hall probe is scanned over the sample surface to measure the perpendicular component of the surface magnetic fields using conventional Scanning Tunneling Microscopy (STM) positioning techniques as shown in Fig. 1. The SHPM system can be designed to enable operation over a wide temperature range, (30 mK–300 K). The SHPM has started as an obscure SPM method with very limited performance and applications at cryogenic temperatures, to a well-fledged magnetic characterization technique with 50 nm spatial resolution and extremely high field resolution. The first microscope [1] could get an image over many hours at 4 K due to noise; in contrast to 1 frame/s scan speeds of modern SHPMs, even at room temperature. The state of the art SHPMs' field resolution is now capable of imaging magnetic vortices in room temperature superconductors, if one day, someone discovers them. SHPM is a quantitative and non-invasive method, compared to MFM with similar spatial resolution. The nano-Hall sensor's used in the SHPMs do not saturate under high external fields and the technique has been shown to operate up to 16 T. The microscope can also be used as a local scanning magnetometer with extremely high magnetic moment sensitivity and can measure magnetization loops of individual magnetic nanodots. SHPMs have recently been commercialised for low and room temperature applications [6]. The performance of the SHPMs can be improved further down to 10–20 nm spatial resolution, ps time resolution, 6–8 frames/s scan speeds, without sacrificing the field resolution drastically.

2 Experiment

2.1 Scanning Hall Probe Microscope

The RT-SHPM [6] used in our studies has a scan range of 56 × 56 μm in XY directions and 4.8 μm in Z direction. The SHPM incorporates XYZ motors for coarse micro-positioning, a video camera for Hall sensor alignment, another video microscope integrated into the piezo scanner with ×14 optical magnification for visualization of the sample and an integrated coil concentric to the Hall sensor head for the application of external magnetic fields of up to ±40 Oe. Furthermore, a newly developed compact sized powerful pulse coil can be coupled with the system to apply external fields up to ±25,000 Oe. The Hall sensor is positioned close to a gold-coated corner of a deep etch mesa, which serves as STM tip. The Hall probe chip is tilted ∼ 1° with respect to sample ensuring that the corner of the mesa is the highest point. The microscope can be run in two modes: STM tracking and lift-off mode. In the STM tracking mode, the tunnel current between the corner of the Hall sensor chip and the sample is measured and used to drive the feedback loop enabling the simultaneous measurement of both STM topography and the magnetic field distribution of the sample surface. This mode of operation gives the highest sensitivity because of the smallest probe-sample separation at all times, but with the drawback of being slow. In the lift-off mode, the Hall sensor is lifted off to a certain height above the sample and the head can be scanned extremely fast (∼ 4 s/frame) for measurements of the local magnetic field distribution. AFM tracking SHPM has recently been developed, integrating a micro-Hall sensor onto a SiN AFM cantilever [7] and onto a GaAs AFM cantilever with sharp tip [8]. The force is measured and controlled by optical [7] or piezoresistive [8] detection.

The LT-SHPM [6] used in this study is very compact (23.6 mm OD) and has the same features of RT-SHPM. It can operate between 30 mK–300 K and tested up to 16 T external fields. The sample can be positioned within 3 mm in XY directions.

The minimum detectable magnetic field is limited mainly by the noise of the Hall sensor, dominated by Johnson and 1/f noise. Nano-Hall sensors are driven with a DC current, (I_{HALL}) and the Hall voltage measured using a low noise amplifier positioned close to the nano-Hall probe. The amplifier's gain and bandwidth are adjustable parameters. The minimum detectable magnetic field with a Hall probe can be written as [2],

$$B_{min} = V_{noise}/(R_H I_{HALL}) \quad (1)$$

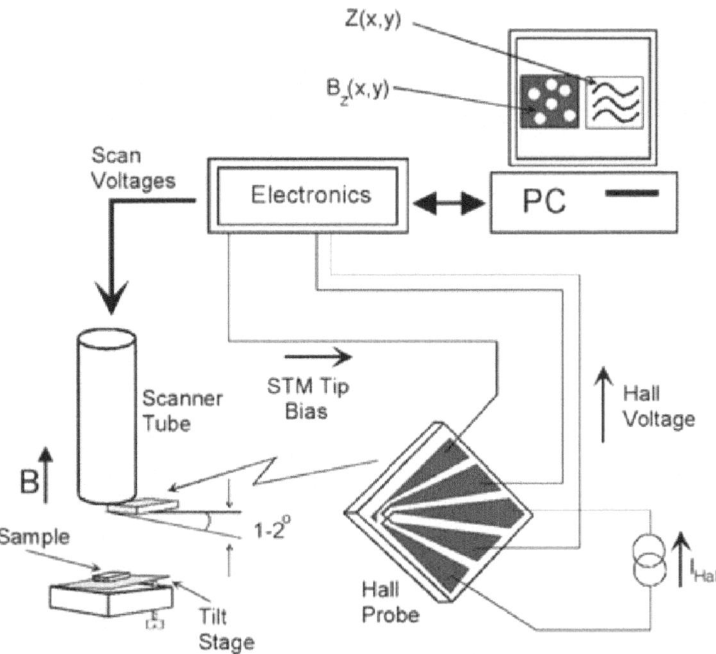

Fig. 1. Schematic diagram of Room Temperature Scanning Hall Probe Microscope (RT-SHPM).

where V_{noise} is the total voltage noise at the input of the Hall amplifier. The voltage noise of the amplifier can usually be made negligible. The V_{noise} has two components; and the Johnson noise due to the series resistance of the Hall sensor (R_s) and the 1/f noise. It is desirable to drive the Hall probe with the highest permissible current. However, the voltage noise of series resistance R_s increases due to heating of the charge carriers and the lattice. Therefore, the Hall current cannot be increased indefinitely and there is a maximum useable $I_{HALLmax}$.

2.2 Hall Probe Fabrication

We earlier exploited excellent properties of GaAs/GaAlAs two Dimensional Electron Gas (2DEG) Hall probes for cryogenic and room temperature measurements [9, 10]. In an attempt to overcome Hall sensor dimension and drive current limitations due to carrier depletion effects in sub-micron GaAs/AlGaAs 2DEG probes at room temperature, we recently fabricated bismuth (Bi) nano-Hall and InSb micro-Hall sensors. Materials with high mobility and low carrier concentrations are desirable for optimal performance. InSb has the highest mobility at room temperature. Combination of optical lithography and focused ion beam milling is used for Hall probe fabrication. The

InSb micro-Hall probes were fabricated on high quality epitaxial InSb thin films with a thickness of 1 μm grown by MBE on semi insulating GaAs substrate [11]. InSb films have a carrier concentration of 2×10^{12} cm^{-2} and a Hall mobility of 55,500 cm^2/Vs. The process used for the micro-fabrication of the InSb Hall sensors was similar to the GaAs 2DEG sensors as reported previously [2]. Figure 2(a) shows a $\sim$ 1.5 μm size InSb micro-Hall probe with a Hall coefficient of $R_H \sim 0.034$ Ω/Gauss and a series resistance of $R_s = 2.2k$ Ω. InSb thin film micro-Hall sensors exhibit a noise level of 6–10mG/$\sqrt{\text{Hz}}$, which is an order of magnitude better than GaAs/AlGaAs 2DEG sensors.

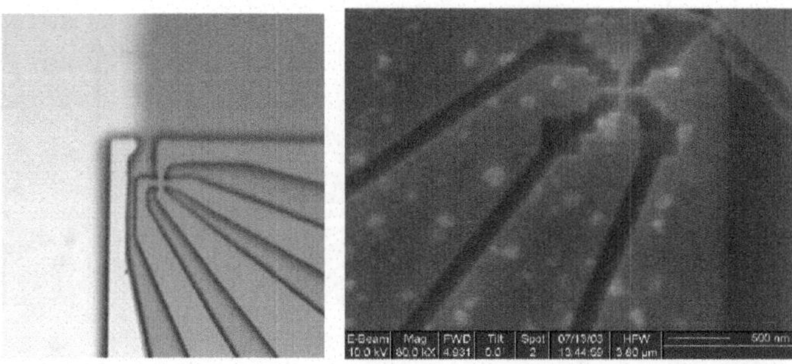

Fig. 2. Optical micrograph of a 1.5 μm InSb thin film micro-Hall (a) and a 50 nm Bismuth nano-Hall probe (b). Gold coated corner of the chip on the left of Hall cross serves as STM tip.

We have earlier achieved a spatial resolution of 120 nm [12] and recently 50 nm [13] using Bi nano-Hall sensors. However, the minimum detectable magnetic field with the Bi nano-Hall sensors was higher, 1 G/$\sqrt{\text{Hz}}$, than InSb due to high carrier concentration and low mobility at the room temperature.

3 Results

3.1 Imaging Hard Disk Media and Tape Head

Figure 3 shows the STM topography (a) and magnetic field image (b) of a CoCrTa Hard Disk sample obtained on NIST calibration sample [14] using STM tracking SHPM scans. An 800 nm GaAs/GaAlAs 2DEG Hall sensor is used for the scan. The SHPM images are quantitative giving magnetic field directly at every pixel as shown in Fig. 3(c).

Figure 4 shows the SHPM images of a sub-micron gap tape head from StorageTek at various write current levels, from 40 mA to −40 mA. Since SHPM is quantitative, it is extremely useful for measuring fields generated by tape heads or hard disk write heads, enabling engineers to design better heads and the media.

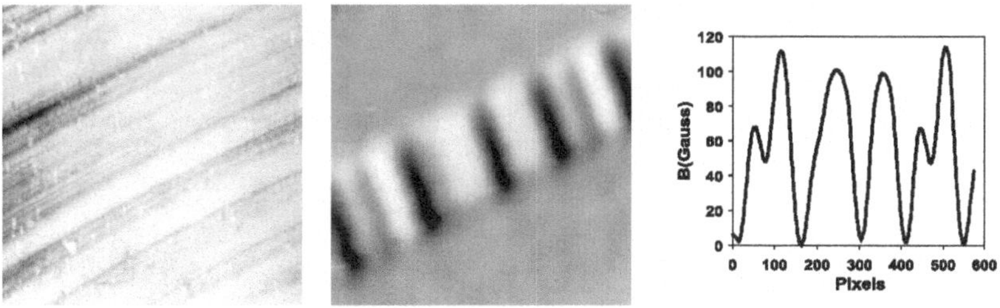

Fig. 3. Simultaneous STM (a) and SHPM image of a data track in CoCrTa Hard Disk Sample obtained in STM tracking SHPM mode. (c) shows the cross section along the arrow.

Fig. 4. 50 μm × 50 μm RT-SHPM images of 800 nm gap StorageTek Tape Head at (a) +40 mA, (b) +20 mA, (c) 0 mA, (d) −20 mA (e) −40 mA drive current. (f) optical microscope image of the tape head array.

3.2 Imaging of Magnetic Materials

Figure 5 shows the magnetic and topography images of a polycrystalline NdFeB sample obtained with the RT-SHPM simultaneously. A 800 nm size Hall sensor microfabricated from a P-HEMT wafer is used in the experiment. The sensor had a 3 mΩ/G Hall coefficient and a 25 μA DC Hall current is passed during the operation of the SHPM.

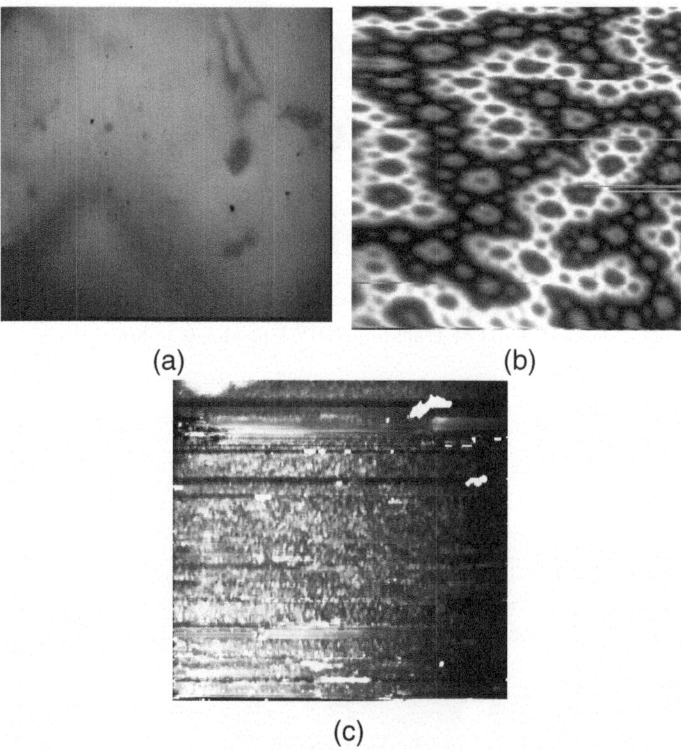

Fig. 5. Sample and the back of the Hall sensor as seen from the video microscope (a), SHPM image of the NdFeB sample, 56×56 μm scan area, black to white corresponds to -2300 to $+2400$ Gauss (b) and simultaneous STM topography of the sample, black to white corresponds to 300 nm.

3.3 Imaging of Superconductors and Magnetization Measurements

Magnetic field penetrates into superconductors as quantised fluxons called Abrikosov vortices. We have imaged vortices in YBCO and BSCCO superconductors using our LT-SHPMs. Figure 6 shows the image of vortices forming a regular triangular lattice in BSCCO single crystal [15]. SHPMs can also be used to perform local magnetization measurements, over an area defined by the Hall probe size, down to 50 nm scale. The Hall sensors are probably the smallest magnetometers that can be constructed with extremely high magnetic moment sensitivity. Moreover, one can scan the sample and measure the magnetization over an area, operating it as a Scanning Hall Magnetometer. We have also performed quasi-real time imaging to investigate how vortices penetrate into the superconductor. Figure 6 shows snapshots of images acquired at 1s intervals, showing the motion of vortices as they penetrate the

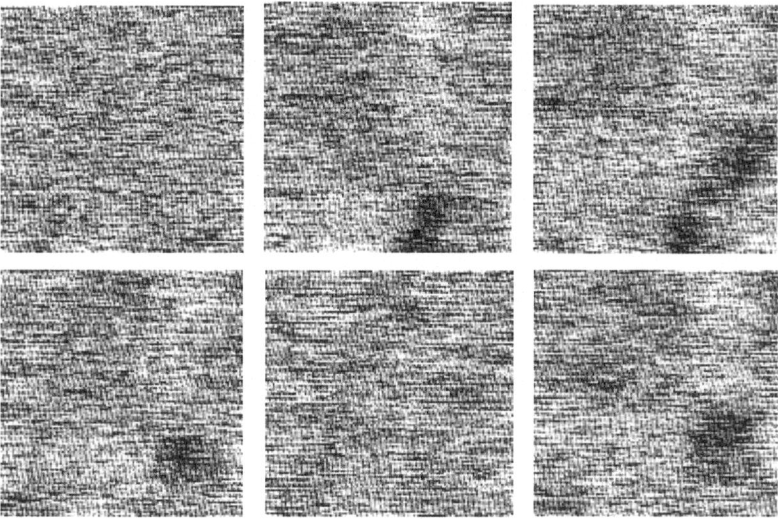

Fig. 6. Snapshots of SHPM images showing the penetration of individual vortices into BSCCO single crystal at 77 K as the Hext is cranked up to +4 Oe from 0.

crystal one by one. They stop at the pinning sites momentarily as they are move under the influence of the external field. One can make movies out of these snapshots to visualize the events more clearly.

4 Conclusion

SHPMs are becoming indispensable tools for characterizing the magnetic materials at ($\sim$ 50 nm) scale quantitatively and non-invasively. Improvements in the spatial resolution ($\sim$ 10–20 nm), time resolution ($\sim$ ps) and scan speed ($\sim$ 8–10 Frames/s) seem to be feasible, opening up new possibilities, which can not be imagined before.

Acknowledgments

Contributions of M. Dede, M. Demir, M. Özdag and M. Kaval are gratefully acknowledged. We thank Dr. J. McCord for supplying the NdFeB sample, Prof. M. Parlak for Bismuth film growth, Prof. K. Kadowaki for BSCCO single crystal sample and Dr. Hadas Shtrikman for 2DEG materials. This project is partially funded by TÜBITAK Project No: TBAG-1878, NanoMagnetics Instruments Ltd. and Turkish Academy of Sciences, TÜBA.

References

1. A.M.Chang, H.D.Hallen, L.Harriot, H.F.Hess, H.L.Loa, J.Kao, R.E.Miller, and T.Y.Chang, Appl. Phys. Lett. 61,1974 (1992).
2. A. Oral, S. J. Bending and M. Henini, "Real-time Scanning Hall Probe Microscopy", Appl. Phys. Lett., vol. 69, no. 9, pp. 1324–1326, August 1996.
3. Y. Martin and H.K. Wickramasinghe, "Magnetic Imaging by Force Microscopy with 1000Å Resolution", Appl. Phys. Lett., vol. 50, no. 20 pp. 1455–1457, May 1987.
4. E. Betzig, J.K. Trautman, R. Wolfe, E.M. Gyorgy, P.L. Finn, M.H.Kryder, C.H.Chang "Near-Field Magnetooptics And High-Density Data-Storage", Appl. Phys. Lett., vol. 61, no. 2, pp. 142–144, July 1992.
5. F. Schmidt and A. Hubert, "Domain Observations on CoCr-Layers With a Digitally Enhanced Kerr-Microscope", J. Mag. Magn. Mat. vol. 61, no. 3, pp. 307–320, October 1986.
6. Low Temperature Scanning Hall Probe Microscope (LT-SHPM) and Room Temperature Scanning Hall Probe Microscope (RT-SHPM), NanoMagnetics Instruments Ltd. 17 Croft Road, Oxford, U.K. www.nanomagnetics-inst.com
7. B.K Chong, H. Zhou, G. Mills, L. Donaldson, J.M.R. Weaver, J.Vac. Sci. & Tech. A 19 (4): 1769–1772 (2001)
8. A. J. Brook, S. J. Bending, J. Pinto, A. Oral, D. Ritchie, H. Beere, A. Springthorpe, and M. Henini, J. Micromech. Microeng. 13(1), 124–128, 2003
9. A. Oral, S. J. Bending and M. Henini, "Scanning Hall Probe Microscopy of Superconductors and Magnetic Materials", J. Vac. Sci. & Technol. B., vol. 14, no. 2, pp. 1202–1205, March–April 1996.
10. A. Sandhu, H. Masuda, A. Oral and S.J. Bending, "Direct Magnetic Imaging of Ferromagnetic Domain Structures by Room Temperature Scanning Hall Probe Microscopy Using a Bismuth Micro-Hall Probe", Jpn. J. Appl. Phys. vol. 40, no. 5B Part 2, pp. L524–L527, May 2001.
11. A. Oral, M. Kaval, M. Dede, H. Masuda, A. Okamoto, I. Shibasaki and A. Sandhu, IEEE Transactions on Magnetics. 38 (5), 2438–2440 (2002)
12. A. Sandhu, H. Masuda, K. Kurosawa K, A. Oral and S.J. Bending, "Bismuth nano-Hall probes fabricated by focused ion beam milling for direct magnetic imaging by room temperature scanning Hall probe microscopy", Elec. Lett., vol. 37, no. 22, pp. 1335–1336, October 2001.
13. A. Sandhu, K. Kurosawa , M. Dede and A. Oral Jap. J. Appl. Phys. 43, 777 (2004).
14. G.D. Howells, A. Oral, S.J. Bending, S.R. Andrews, P.T Squire, P. Rice, A. de Lozanne, J.A.C. Bland, I. Kaya and M. Henini , J. Magn. and Magn. Mat., 197, 917–919 (1999)
15. A.Oral, J.C.Barnard, S.J.Bending, I.I.Kaya, S.Ooi, H.Taoka, T.Tamegai and M.Henini., Phys. Rev. Lett. 80, 3610–3613 (1998).

Self-Assembled FePt Nanoparticle Arrays as Potential High-Density Recording Media

Shouheng Sun

IBM T. J. Watson Research Center, Yorktown Heights, New York 10598, USA

Summary. In this chapter recent synthetic progress in FePt and CoPt nanoparticles and nanoparticle arrays are summarized. The importance of monodisperse nanoparticles for future ultrahigh density data storage is mentioned. General chemical process for making magnetic nanoparticles and nanoparticle arrays is described. The synthetic progress that has been made so far and the problems indentified are summarize. Finally, possible future direction in the synthesis for practical high density recording applications is pointed out.

1 Introduction

Increasing the recording density of a computer hard disk has spurred tremendous interests in the research and development community of data storage due to the scientific and technological challenges facing such increase. Present hard disk is written and read longitudinally and the magnetizations of the recorded bits lie in the disk plane. The recording system contains a head with a separate read and write element, as illustrated in Fig. 1A [1, 2]. The head flies in close proximity to a granular recording medium to reduce the bit dimension and increase the recording density. Recording media traditionally have a single magnetic storage layer and consist of weakly coupled magnetic grains, or particles, of CoPtCrX alloy (X = B, Ta), as shown in Fig. 1B. The fine microstructure of the grains allows for smaller bits of magnetic transitions (Fig. 1A and B) and narrower gap between the two transitions (the inset of Fig. 1A), and therefore the higher recording density.

Advances to high density of magnetic recording have resulted primarily from proportional scaling of the recording apparatus, including write/read head, media thickness, and grain size. As the signal-to-noise ratio of the medium depends on the number of magnetic grains within each bit, maintaining this ratio at an acceptable level requires the development of smaller magnetically stable grains with high coercivity, low magnetization, and minimal magnetic exchange coupling between the neighboring grains [3]–[8]. To achieve these goals, media based on monodisperse magnetic nanoparticle arrays have been proposed [9, 10]. The particles coated with non-magnetic layer are pre-synthesized with controlled size and size distribution and are then assembled on a solid substrate via a self-assembly process. By controlling

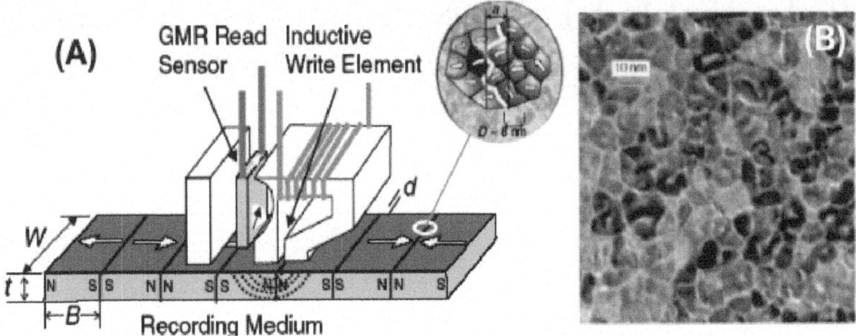

Fig. 1. (A) Schematic drawing of a longitudinal recording system in which B is the bit length, W is the track width, t is the medium thickness and d is the flying height of the head above the medium. The inset schematically shows a magnetic transition gap between two magnetic transitions. (B) TEM image of the modern recording CoCrPtB media.

deposition conditions, these nanoparticles can form regularly arrayed structure with characteristic dimensions much smaller than those conceivable with physical deposition and lithographic methods. Such an array with controlled magnetics is able to support magnetic recording at the density well beyond the Terabits/in^2 regime.

2 FePt Nanoparticles as Media Materials

The push to higher and higher magnetic recording density has led to the reduction of the average size of current cobalt-based magnetic grains to about 8–10 nm, a dimension close to the onset of its superparamagnetic behavior – the thermal energy is comparable with the magnetic anisotropy energy, resulting in magnetic signal decay and loss of recorded information. Therefore, in a media containing smaller magnetic grains, high K_u material is needed to maintain the high magnetic energy K_uV, and thus bit stability, against the thermal agitation. A minimal stability ratio of stored magnetic energy to thermal energy, K_uV/k_BT is believed to be 50–70 [1]. FePt material at the composition close to $Fe_{50}Pt_{50}$ is a promising candidate for this high density media application. It can form an intermetallic compound with a chemically ordered face centered tetragonal (fct) structure in which the Fe and Pt atoms form alternative layers along the c direction, its magnetic easy axis direction (Fig. 2). This fct structured FePt is known to have very high magnetic anisotropy constant K_u, reaching 10^8 erg/cm^3, a value that is about 50–100 times larger than that in the CoPtCrX alloys used for the current advanced media. This large K_u offers stable minimal grains down to sizes of 2.8 nm, about a factor of 4 smaller than the grain size in typical 100 Gbit/in^2 media [1].

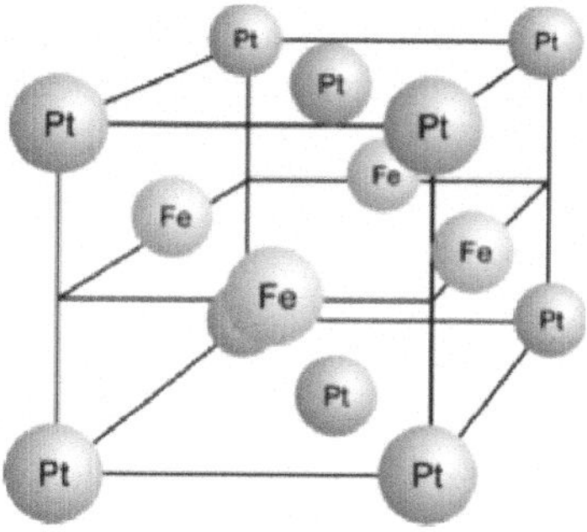

Fig. 2. A unit cell of fct structured FePt.

3 FePt Nanoparticle Synthesis

Small magnetic FePt nanoparticles are commonly fabricated using vacuum deposition techniques [11]–[14]. As-deposited, the FePt has chemically disordered fcc structure and is magnetically soft. Thermal annealing is needed to transform the fcc structure into the chemically ordered fct structure. However, the annealing also results in particle aggregation and strong exchange coupling among the particles. To reduce the particle size and eliminate the exchange-coupling, small FePt particles prepared from vacuum deposition methods are often buried into a variety of insulator matrixes of SiO_2 [15] Al_2O_3 [16, 17], B_2O_3 [18] or S_3N_4 [19]. Alternatively, FePt particles can be made via gas-phase evaporation [20]. Although the average particle size can be better controlled during the high temperature annealing, it is still difficult in using these particles to form regular arrays required for future recording applications. Different from all the physical deposition processes, solution phase synthesis offers a unique way of making monodisperse magnetic FePt nanoparticles. In solution, the particles are first grown in a homogenous nucleation step, followed by an isotropic growth of the nuclei. The small particles are stabilized by organic surfactants, and stable FePt nanoparticle dispersions in various solvents can be readily made. In the synthesis, reaction conditions and the chemical nature of the organic surfactants are often used to tune the size, shape and composition of the particles. The monodisperse FePt nanoparticles are commonly synthesized by the decomposition of iron pentacarbonyl, $Fe(CO)_5$, and reduction of platinum acetyacetonate, $Pt(acac)_2$ in a high boiling solvent [21]–[28]. The decomposition of $Fe(CO)_5$ at higher temperature gives Fe-containing species that can quickly combine with the Pt-containing species formed from the reduction of Pt salt, yielding FePt nuclei. The particles grow via the coating of the nuclei with more FePt from the precursors and the growth is terminated by the reaction conditions and surfactants used in the solution. Alternatively, the particles can be made via co-reduction of $Pt(acac)_2$ and iron salt with various reducing agents [29]–[33]. The oleic acid and oleylamine are two common organic surfactants used for FePt particle protection. Oleic acid reacts with the surface Fe to form -COO-Fe and oleylamine coordinates to the surface Pt, giving -NH_2-Pt. These surface reactions result in a robust organic coating layer surrounding each FePt particle, preventing them from agglomeration. The reaction is usually conducted in high temperature solution phase (260–300°C) to facilitate the decomposition and reduction reactions and the formation of the FePt particles. The composition of the particles is controlled by the molar ratio of both Fe and Pt precursors. The size of the particles can be tuned from 2 to 10 nm. The particles stabilized with oleic acid and oleylamine can be easily dispersed into various hydrocarbon solvents, facilitating their self-assembly into FePt nanoparticle superlattices.

4 Self-Assembly of FePt Nanoparticles

Nanoparticles coated with organic surfactants can form close-packed arrays on a variety of substrates as the solvent from the particle dispersion is allowed to evaporate. This process is called nanoparticle self-assembly. In this self-assembled structure, the nanoparticles are the building blocks and are connected by van der Waals and magnetic dipole interactions. FePt nanoparticles prepared from solution phase chemistry and stabilized with oleic acid and oleylamine are ideal building blocks for constructing FePt nanoparticle superlattices. Two assembly examples are shown in Fig. 3 [21]. In Fig. 3A, the superlattice of 6 nm FePt nanoparticles is obtained from a drop ($\sim$ 0.5 μL) of the dilute hexane/octane (v/v $\sim$ 1/1) dispersion ($\sim$ 1 mg/mL) of the particles. The interparticle spacing in this hexagonal close packed 3D array is around 4 nm maintained by the oleic acid and oleylamine capping layers. The original capping layer can be replaced by new capping layers via surface surfactant exchange and thus interparticle spacing and assembly structure can be controlled. Figure 3B shows the self-assembled superlattice of the 6 nm FePt nanoparticles coated with C_5H_{11}-COOH and C_6H_{13}-NH_2 via room temperature surfactant exchange in toluene solution. The assembly exhibits a cubic packed structure. Due to the shorter hydrocarbon chain presented by C_5H_{11}-COOH and C_6H_{13}-NH_2, the interparticle spacing is now reduced to only $\sim$ 1 nm, resulting in higher particle packing density.

5 Thermal Annealing Induced Structural Change in Self-Assembled FePt Nanoparticle Arrays

As the FePt particles are connected by weak interactions, the self-assembled nanoparticle arrays are mechanically soft. To make robust particle assemblies, thermal annealing can be applied to induce chemical reactions among the stabilizers, and the interface reactions between the particles and the substrate.

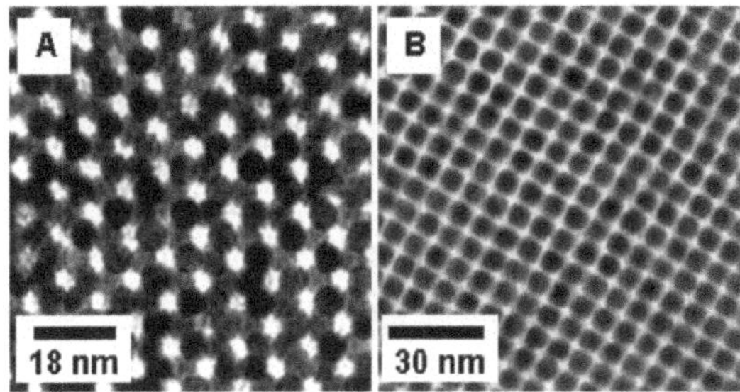

Fig. 3. (A) TEM micrograph of a 3D assembly of 6 nm as-synthesized $Fe_{50}Pt_{50}$ particles deposited from a hexane/octane (v/v 1/1) dispersion. (B) TEM micrograph of a 3D assembly of 6 nm $Fe_{50}Pt_{50}$ sample after replacing oleic acid/oleylamine with hexanoic acid/hexylamine.

The organic surfactants in the hydrocarbon coating layer are not thermally stable and tend to react under a controlled annealing condition to give polymeric hydrocarbon that surrounds each particle, as illustrated in Fig. 4. For a 180 nm thick 4 nm $Fe_{52}Pt_{48}$ particle assembly annealed at 560°C for 30 minutes, the HRSEM images of both surface (Fig. 5A) and cross section (Fig. 5B) of the assembly show that the ordering in the superlattice array of the particles is retained [21]. Interparticle spacings, however, are reduced from ~ 4 to ~ 2 nm due to the decomposition of the hydrocarbon coating. Some coherent strain is also observed in the superlattices due to this shrinkage. Rutherford backscattering measurements on these annealed 4 nm $Fe_{52}Pt_{48}$ particle assemblies indicate 40–50% (at. %) carbon content, indicating that annealing at high temperature does not result in the loss of stabilizing ligands; rather, they are converted to a carbonatious coating around each particle.

Thermal annealing also induces the internal particle structure change from chemically disordered fcc to chemically ordered fct (Fig. 4) [21]. This structure transition can be easily monitored with the wide-angle X-ray diffraction (XRD). Figure 6 shows a series of XRD patterns for ~ 1 mm thick $Fe_{52}Pt_{48}$ assemblies as a function of annealing temperature at a constant annealing time of 30 minutes. The as-synthesized particles exhibit the chemically disordered fcc structure (Fig. 6A). Annealing induces the Fe and Pt atoms to rearrange into the long range chemically ordered fct structure, as indicated by the (111) peak shifts and evolution of the (001), (110) peaks (Fig. 6B–E). At annealing temperatures below 500°C, only partial chemical ordering is observed (Fig. 6B, C). The chemical ordering can be increased by annealing at higher temperatures (Fig. 6D, E) or by increasing the annealing time. XRD [21], Kerr effect [34], and in-situ HRTEM measurements [35] on annealed $Fe_{52}Pt_{48}$ particle assemblies show that the onset of this phase change occurs at 500–530°C. While annealing at higher temperatures or for longer time can increase the chemical ordering, too high a temperature, e.g.

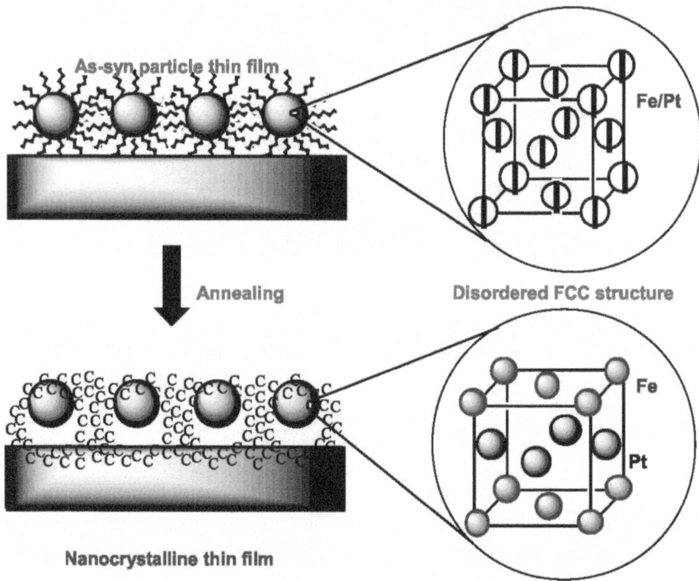

Fig. 4. Schematic illustration of surface and internal structural changes of FePt nanoparticles due to thermal annealing.

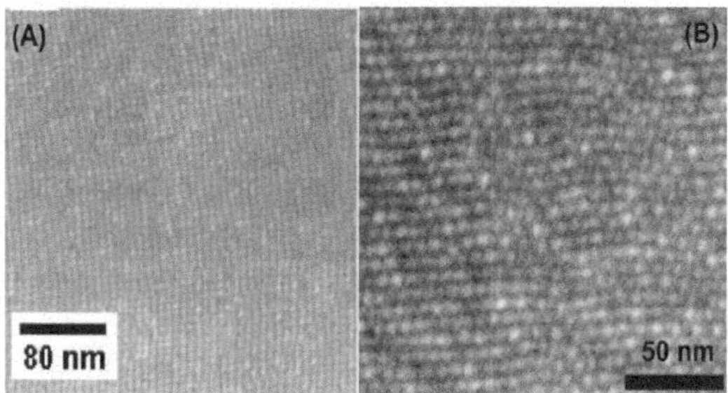

Fig. 5. HRSEM images of (A) a surface view and (B) a part of the side view of 180 nm thick 4 nm $Fe_{52}Pt_{48}$ nanoparticle assembly annealed at 560°C for 30 minutes under 1 atmosphere of static N_2 gas.

$>$ 600°C, or too long an annealing time will result in either aggregation or phase separation of the particles. XRD combined with TEM and HRSEM studies show that the optimum temperature for the phase transformation is 580°C. It is worth mentioning, however, that the exact structure after thermal annealing depends strongly on the particle composition. Figure 7 shows a series of XRD patterns of differently composed FePt assemblies annealed at 580°C for 30 minutes [22]. It shows that among all these 580°C annealed

FePt nanoparticle assemblies, only the $Fe_{56}Pt_{44}$ assembly yields a high quality fct phase, an observation that is consistent with those from the vacuum deposited granular films.

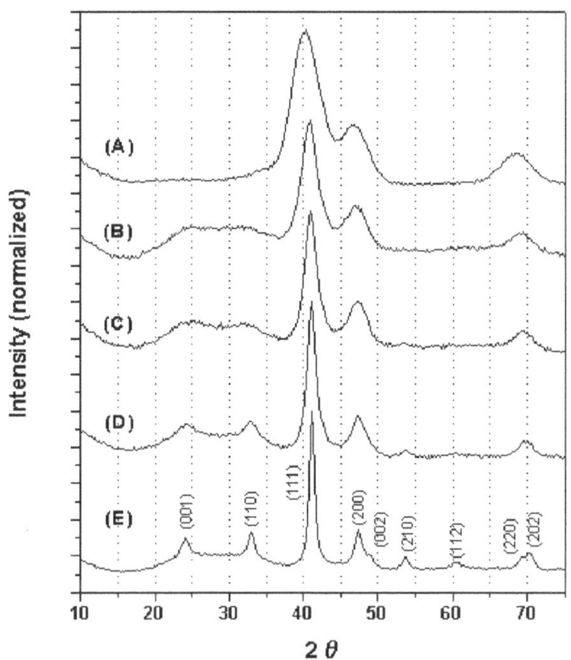

Fig. 6. X-ray diffraction patterns (A) of as synthesized 4 nm $Fe_{52}Pt_{48}$ particle assemblies, and a series of similar assemblies annealed under atmospheric N_2 gas for 30 minutes at temperatures of (B) 450°C, (C) 500°C, (D) 550°C and (E) 600°C. The indexing is based on tabulated fct FePt reflections. The diffraction patterns were collected with a Siemens D-500 diffractometer using Cu Ka radiation (λ = 1.54056 Å).

6 Thermal Annealing Induced Magnetic Property Change in Self-Assembled FePt Nanoparticle Assemblies

The as-synthesized fcc FePt particles are superparamagnetic at room temperature. They are ferromagnetic only at very low temperature. The temperature dependent magnetization was measured in a 10 Oe field between 5 and 400 K using the standard zero-field-cooling and field-cooling procedures. These studies indicate that superparamagnetic behavior is blocked at 20–30 K. Figure 8

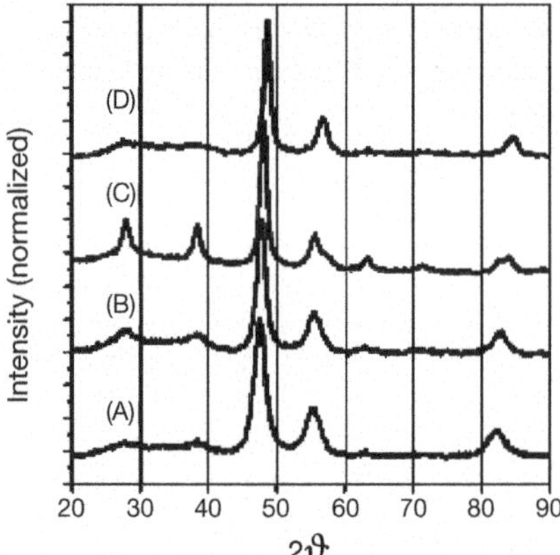

Fig. 7. XRD patterns of (A) $Fe_{38}Pt_{62}$, (B) $Fe_{48}Pt_{52}$, (C) $Fe_{56}Pt_{44}$, and (D) $Fe_{70}Pt_{30}$ nanoparticle assemblies annealed at 580°C for 30 minutes. The diffraction patterns were collected with a Siemens D-500 diffractometer using Co Ka radiation ($\lambda = 1.788965$ Å).

shows temperature dependent H_c of the 4 nm $Fe_{56}Pt_{44}$ nanoparticle assemblies. At 5 K, the assembly shows an H_c of 4000 Oe (Fig. 8A). This coercivity drops sharply as the temperature is raised to 15 K (Fig. 8B), and reaches superparamagnetism at 35 K or above. This low transition temperature between ferromagnetism and superparamagnetism is consistent with the low magnetocrystalline anisotropy of the fcc structure of FePt.

Annealing converts the particle structure from fcc to the high anisotropic fct phase and transforms them into room temperature nanoscale ferromagnets. Room temperature coercivities of the FePt nanoparticle assemblies are tunable by controlling the annealing temperature and time. Figure 9 shows the room temperature hysteresis loops for 4 nm $Fe_{56}Pt_{44}$ particle assemblies that were annealed for 30 minutes at 500, 550, and 580°C, respectively [22]. The H_c values increase dramatically with annealing temperature, showing the transition from superparamagnetic to ferromagnetic behavior. The sample annealed at 500°C appears nearly superparamagnetic at room temperature. An expanded view of the low-field part of the loop shows an HC value of 330 Oe and the loop is hysteretic up to 6 kOe (Fig. 9A). This likely results from a minority fraction of the particles having sufficient anisotropy to be ferromagnetically ordered at room temperature. The sample annealed 550°C increases the ferromagnetic fraction of particles and results in the H_c value increase to 3200 Oe (Fig. 9B). However, there is still an inflection in the magnetization curve near 0 Oe, indicating the existence of low H_c particles. The sample annealed at 580°C show a room temperature H_c of above 9000 Oe with a loop shape typical for isotropic distribution of high anisotropy particles (Fig. 9C). The coercivity of the 4 nm FePt nanoparticle assemblies also

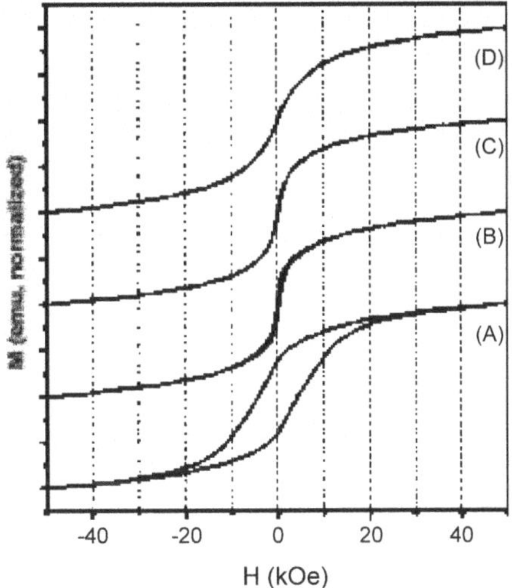

Fig. 8. Low temperature hysteresis loops of the as synthesized 4 nm $Fe_{56}Pt_{44}$ nanoparticle assemblies at (A) 5 K, (B) 15 K, (C) 35 K and (D) 85 K.

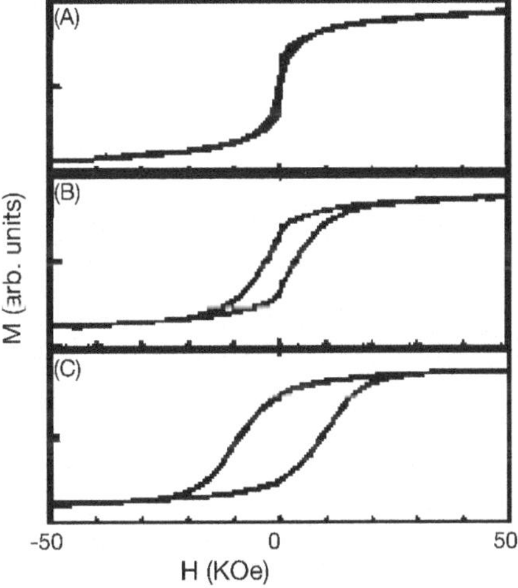

Fig. 9. Room temperature hysteresis loops of 4 nm $Fe_{56}Pt_{44}$ nanoparticle assemblies annealed at (A) 500°C, (B) 550°C, and (C) 580°C.

depends on the composition of the particles. Figure 10 shows the in-plane coercivity data for a series of 140 nm thick 4 nm FePt samples as a function of the composition [22]. Figure 10A shows the 5K data of the as-synthesized FePt nanoparticle assemblies and Fig. 10B is the room temperature data for the assemblies annealed at 580°C for 30 minutes. In both cases, the hysteresis data of the FePt nanoparticle assemblies show similar coercivity dependence on the composition. The extrapolation of the Gaussian type fit yields the highest-H_c composition of FePt to be $Fe_{55}Pt_{45}$, which is consistent with the earlier reports on vacuum deposited FePt thin films.

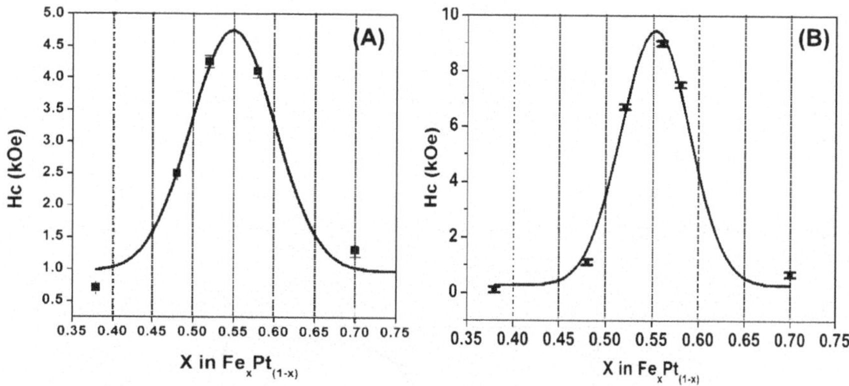

Fig. 10. Composition dependent coercivity H_c of 580°C annealed FePt nanoparticle assemblies.

7 Self-Assembled FePt Nanoparticle Array to Support Magnetization Reversal Transitions

The annealed FePt nanoparticle assemblies are smooth ferromagnetic films and can support magnetization reversal transitions (bits). Atomic force microscopy studies on a 120 nm thick assembly of 4 nm $Fe_{48}Pt_{52}$ particles indicate a 1 nm root-mean-square variation in height over areas of 3×3 mm. This smooth assembly has an in-plane coercivity of 1800 Oe at room temperature, enough to support the magnetization transitions. Figure 11A shows the read-back sensor voltage signals from the written data tracks recorded using a static write/read tester [21]. The individual line scans reveal magnetization reversal transitions at linear densities of 500, 1040, 2140, and 5000 flux changes per millimeter (fc/mm) (curves a–d, respectively). These write/read experiments demonstrate that this 4 nm $Fe_{48}Pt_{52}$ ferromagnetic nanoparticle assembly supports magnetization reversal transitions at reasonable linear densities that can be read back nondestructively. The thermal stability of the transitions was assessed using a dynamic coercivity method which relies on remanent coercivity measurements as a function of the applied magnetic field pulse width. Figure 11B shows respective data, from which the ratio of the energy barrier for magnetization reversal (KuV) to the thermal energy (kBT) is extracted to be KuV/kBT = 48. This corresponds to an average anisotropy constant <Ku> of 5.9×10^7 erg/cm^3.

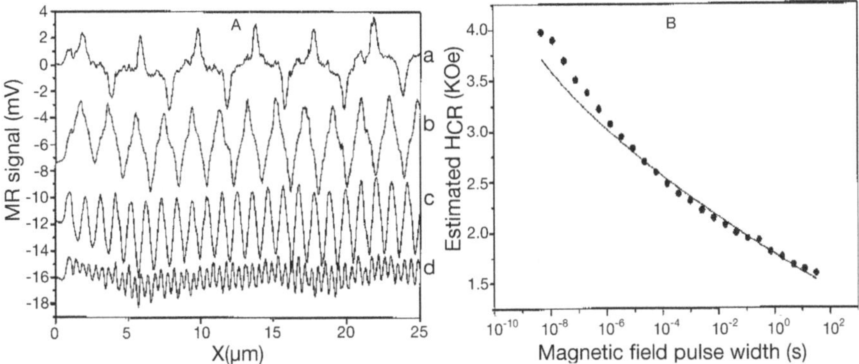

Fig. 11. (A) Magneto-resistive (MR) read-back signals from written bit transitions in a 120-nm-thick assembly of 4-nm-diameter $Fe_{48}Pt_{52}$ nanoparticles. The individual line scans reveal magnetization reversal transitions at linear densities of (a) 500, (b) 1040, (c) 2140, and (d) 5000 fc/mm. (B) Dynamic coercivity measurements (HCR) of the sample in (A) at 300 K over a range from 5 ns to 65 s. The measured data (v) are fit to a dynamic coercivity law for pulse width <10>$^{-6}$ s (solid curve).

For an assembly with larger coercivity, thermally assisted writing can be applied to record the magnetization transitions. For example, a 10 nm thick assembly of 4 nm $Fe_{58}Pt_{42}$ nanoparticles that was annealed at 530°C under Ar + H_2 (5%) for 30 minutes shows an H_c over 5000 Oe at room temperature [36]. Due to this large coercivity, the magnetization of the assembly can not be saturated or reversed using the normal write/read tester. But the magnetization directions can be aligned using the thermally assisted writing technique. A sharply focused laser diode beam with fast pulses ($<$ 100 ns) can induce the local temperature increase to over 200°C. The coercivity of the nanoparticles is reduced at these temperatures and allows for thermally assisted magnetization reversal of the particles by a relatively weak magnetic bias field. Figure 12A shows the AFM image and Fig. 12B is the magnetic force microscope (MFM) image of the 3-layer 4 nm $Fe_{58}Pt_{42}$ assembly that has been treated with a focused pulsing laser (93 ns) under a perpendicular magnetic field (2.5 kOe) [36]. The AFM image shows that the smooth FePt nanoparticle assembly is intact after the laser treatment. The black spots in Fig. 12B indicate the magnetization pointing to out of particle assembly plane. It was also found that the reduced laser power resulted in smaller bits, suggesting that it is possible to use pulsing laser to write magnetic patterns in various packing densities on a thin FePt nanoparticle assembly.

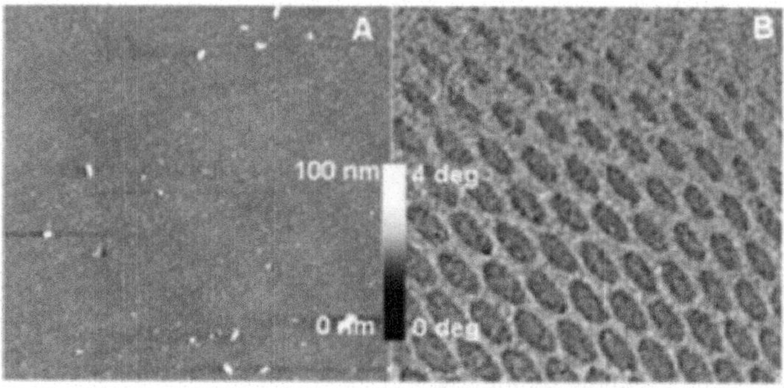

Fig. 12. (A) AFM image (17.4×17.4 mm^2) of a 3-layer 4 nm $Fe_{58}Pt_{42}$ nanoparticle assembly annealed at 530°C under Ar + H_2 (5%) for 30 minutes and treated with a focused pulsing laser (93 ns) under a perpendicular magnetic bias field (2.5 kOe). (B) MFM image of the film in (A).

8 Self-Assembled FePt Nanoparticle Array as Potential High Density Recording Media

Today, hexagonal Co-alloys are widely adopted for media applications because of their high uniaxial magnetocrystalline anisotropies. Thermal stability limitations, however, make it unlikely that the grain sizes of the Co-alloys can be further reduced for ultrahigh density requirement. Self-assembled FePt nanoparticle assembly can be made thinner with higher coercivity and will permit drastically reduced bit cell sizes and as a result, the increased recording density. For example, at 1 Tbit/in^2 areal density, one would still maintain 40 4 nm particles per bit cell with the size of 25.4 nm $\times$ 25.4 nm and bit-aspect-ratio of 1, as opposed to only about 4 grains per bit cell in a more conventional approach. If only considering the thermal stability alone and assuming each FePt nanoparticle can be used to support magnetization transition, one can get a density up to 60 Tbit/in^2 in a self-assembled regular array of ferromagnetic 6 nm FePt nanoparticles with interparticle spacing smaller than 2 nm [9]. To achieve this lofty goal needs, of course, the solutions for the expected problems, including large-scale disk coating capability, disk roughness control, and magnetic easy axis orientation control. These require continued research efforts in producing self-assembled superlattice structures with more precise control on magnetics. Shape controlled synthesis and shape induced texture of the nanoparticles may become a more promising approach to the orientation control on both crystal and magnetic easy axis of the nanoparticles [37].

References

1. D. Weller, A. Moser, IEEE Trans. Magn. **35**, 4423 (1999).
2. A. Moser, et al, J. Phys. D: Appl. Phys. **35**, R157 (2002).
3. E. S. Murdock, R. F. Simmons, R. Davison, IEEE Trans. Mag. **28**, 3078 (1992).
4. T. Yogi, T. A. Nguyen, IEEE Trans. Mag. **29**, 307 (1993).
5. P.-L. Lu, S. H. Charap, IEEE Trans. Mag. **30**, 4230 (1994).
6. D. N. Lambeth, E. M. T. Velu, G. H. Bellesis, L. Lee, D. E. Laughlin, J. Appl. Phys. **79**, 4496 (1996).
7. J. Li, M. Mirzamaani, X. Bian, M. Doerner, S. Duan, K. Tang, M. Toney, T. Arnoldussen, M. Madison, J. Appl. Phys. **85**, 4286 (1999).
8. D. A. Thompson, J. S. Best, IBM J. Res. Develop. **44**, 311 (2000).
9. S. Sun, D. Weller, C. B. Murray: in 'The Physics of High Density Magnetic Recording', Eds. M. Plumer, J. van Ek, D. Weller, Springer-Verlag, Chapter 9 (2001).
10. S. Sun, D. Weller, J. Mag. Soc. Jp. **25**, 1434 (2001).
11. A. Cebollada, D. Weller, J. Sticht, G. R. Harp, R. F. C. Farrow, R. F. Marks, R. Savoy, J. C. Scott., Phys. Rev. B **50**, 3419 (1994).
12. R. F. C. Farrow, D. Weller, R. F. Marks, M. F. Toney, A. Cebollada, G. R. Harp, J. Appl. Phys. **79**, 5967 (1996).
13. Y. Ide, T. Goto, K. Kikuchi, K. Watanabe, J. Onagawa, H. Yoshida, J. M. Cadogan, J. Magn. Magn. Mater. **245**, 177 (1998).
14. R. A. Ristau, K. Barmak, L. H. Lewis, K. R. Coffey, J. K. Howard, J. Appl. Phys. **86**, 4527 (1999).
15. C. P. Luo, S. H. Liou, D. J. Sellmyer, J. Appl. Phys. **87**, 6941 (2000).
16. B. Bian, K. Sato, Y. Hirotsu, A. Makino, Appl. Phys. Lett. **75**, 3686 (1999).
17. B. Bian, D. E. Laughlin, K. Sato, Y. Hirotsu, J. Appl. Phys. **87**, 6962 (2000).
18. C. P. Luo, S. H. Liou, L. Gao, Y. Liu, D. J. Sellmyer, Appl. Phys. Lett. **77**, 2225 (2000).
19. C.-M. Kuo, P. C. Kuo, J. Appl. Phys. **87**, 419 (2000).
20. S. Stappert, B. Rellinghaus, M. Acet, and E. F. Wassermann, J. Cryst. Growth **252**, 440 (2003).
21. S. Sun, C. B. Murray, D. Weller, L. Folks, A. Moser, Science **287**, 1989 (2000).
22. S. Sun, E. E. Fullerton, D. Weller, C. B. Murray, IEEE Trans. Magn. **37**, 1239 (2001).
23. M. Chen, D. E. Nikles, Nano Lett. **2**, 211 (2002).
24. S. Kang, J. W. Harrell, D. E. Nikles, Nano Lett. **2**, 1033 (2002).
25. B. Stahl, N. S. Gajbhiye, G. Wilde, D. Kramer, J. Ellrich, M. Ghafari, H. Hahn, H. Gleiter, J. Weimller, R. Wrschum, P. Schlossmacher, Adv. Mater. **14**, 24 (2002).
26. E. Shevchenko, D. Talapin, A. Kornowski, J. Ktzler, M. Haase, A. Rogach, H. Weller, Adv. Mater. **12**, 287 (2002).

27. B. Stahl, J. Ellrich, R. Theissmann, M. Ghafari, S. Bhattacharya, H. Hahn, N. S. Gajbhiye, D. Kramer, R. N. Viswanath, J. Weissmller, H. Gleiter, Phys. Rev B **67**, 14422 (2003).
28. M. Chen, J. P. Liu, S. Sun, J. Am. Chem. Soc. **126**, 8394 (2004).
29. S. Sun, S. Anders, T. Thomson, J. E. E. Baglin, M. F. Toney, H. F. Hamann, C. B. Murray, B. D. Terris, J. Phys. Chem. B **107**, 5419 (2003).
30. B. Jayadevan, A. Hobo, K. Urakawa, C. N. Chinnasamy, K. Shinoda, K. Tohji, J. Appl. Phys. **93**, 7574 (2003).
31. B. Jeyadevan, K. Urakawa, A. Hobo, N. Chinnasamy, K. Shinoda, K. Tohji, D. D. J. Djayaprawira, M. Tsunoda, M. Takahashi, Jpn. J. Appl. Phys. **42**, L350 (2003).
32. T. Iwaki, Y. Kakihara, T. Toda, M. Abdullah, K. Okuyama, J. Appl. Phys. **94**, 6807 (2003).
33. M. Nakaya, Y. Tsuchiya, K. Ito, Y. Oumi, T. Sano, T. Teranishi, Chem. Lett. **33**, 130 (2004).
34. D. Weller, S. Sun, C. B. Murray, L. Folks, A. Moser, IEEE Trans. Magn. **37**. 2185 (2001).
35. Z. R. Dai, S. Sun, Z. L. Wang, Nano Lett. **1**, 443 (2001).
36. S. Sun, S. Anders, H. F. Hamann, J.-U. Thiele, J. E. E. Baglin, T. Thomson, E. E. Fullerton, C. B. Murray, and B. D. Terris, J. Am. Chem. Soc. **124**, 2884 (2002).
37. H. Zeng, P. M. Rice, S. X. Wang, S. Sun, J. Am. Chem. Soc. **126**, 11458 (2004).

Magnetophotonic Crystals

M. Inoue[1–3], A. Granovsky[2], O. Aktsipetrov[2], H. Uchida[1], and K. Nishimura[1]

[1] Dept. of Electrical & Electronic Engineering, Toyohashi University of Technology, Toyohashi 441-8580, Japan
[2] CREST, Japan Science and Technology Corporation (JST), Kawaguchi 332-0012, Japan
[3] Moscow State University, Moscow 119992, Russia

Summary. Photonic crystals, which are periodic composites of macroscopic dielectric media of different refractive index, affect the propagation of light in much the same way that semiconductor crystals affect the propagation of electrons. When photonic crystals are composed of magnetic materials such as rare-earth iron garnet, they are called as magnetophotonic crystals providing unique optical and magneto-optical properties. For instance, introduction of the magnetic material into the medium as a defect yields very large magneto-optical Faraday effect due to the localization of light at the magnetic defect. In this article, our recent investigations on one-dimensional magnetophotonic crystals composed of rare-earth iron garnet films such as Bi-substituted yttrium iron garnet (Bi:YIG) are summarized, and the very unique properties of the media are demonstrated for the films with $(SiO_2/Ta_2O_5)^k$/Bi:YIG/$(Ta_2O_5/SiO_2)^k$ structures.

1 Introduction

Within the past several years, a growing interest in photonic crystals [1, 2, 3, 4, 5] (one-, two-, or three-dimensional periodic structures of dielectric materials) has resulted from their unique photonic bandgaps (an optical analogy to electron bandgap), where the existence of light (photons) is strictly forbidden. Since the photonic bandgaps yield a novel mechanism for molding the flow of light, the photonic crystals are promised as useful media for opto-electronic devices of next generation.

Our recent calculations [6, 7] have predicted that, when the one-dimensional (1-D) photonic crystals are composed of magnetic materials (magnetophotonic crystals), they exhibit remarkable magneto-optical (MO) properties accompanied by a large enhancement in their Kerr and Faraday rotations. The unique properties arise from the localization effect of light as a result of multiple interference of light within the magnetic multilayers.

In this article, our recent investigations [6, 7, 8, 9, 10, 11, 12, 13] on the magnetophotonic crystals are summarized especially for the case of media with rare-earth iron garnet, and their capabilities as a new class of magneto-optical materials are discussed, both from the theoretical and experimental points of view.

2 Localization of Light and MO Effect

Unique enhancement of the MO effects were first found theoretically for Bi-substituted yttrium iron garnet (Bi:YIG) films with disordered multilayer structures. As shown in Fig. 1, let us consider a multilayer film composed of N_M layers of Bi:YIG and NS layers of SiO_2 which are piled up in an arbitrary sequence. Each Bi:YIG and SiO_2 layers have thicknesses of d_M and d_S, respectively, and the total thickness of the multilayer film is D. The film structure is designed by two structural parameters: one is a binary number b^N with N digit; $b^N = 101110\cdots01$, for instance. Each digit of b^N corresponds to each layer of the film, and "1" and "0" are assigned to the Bi:YIG and SiO_2 layers, respectively. Another parameter is the density of Bi:YIG in the film, P_M, which is defined by $P_M = N_M d_M / D$. For simplicity, all Bi:YIG (SiO_2) layers are assumed to have the same material parameters.

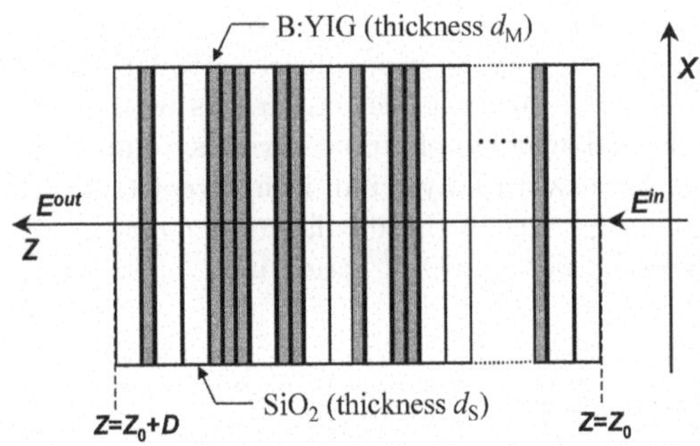

Fig. 1. Schematic drawing of the multilayer film composed of Bi:YIG and SiO_2 layers.

Since the analytical procedure is described elsewhere in detal [6, 7], we confine ourselves to describe the outline of the theoretical results. A typical example of transmissivty T, Faraday rotation angle θ_F, and total rotation angle $\theta = \theta_F N_M d_M$ of disordered multilayer films is shown in Fig. 2 as a function of the Bi:YIG density P_M, where the film is composed of 16 layers with $b^{16} = (\mathrm{ABAA})_{\mathrm{hex}}$ structure. As marked by arrows in Fig. 2, films with (1) $P_M = 0.09$, (2) 0.123, (3) 0.291, and (4) 0.668, show considerably high transmissivity T and large θ_F simultaneously: for instance, T of the film with $P_M = 0.123$ is 91 % and its θ_F is about four times larger than that of the Bi:YIG single-layer film ($\theta_F = -0.1$ deg/μm at $\lambda = 1.15$ μm).

In Figs. 3(a) and 3(b), two cases of the field distribution of light are depicted for films with $P_M = 0.123$ and 0.668, respectively. In both cases, light is weakly localized within the films, suggesting that the unique properties of films originate from the weak localization of light caused by the multiple

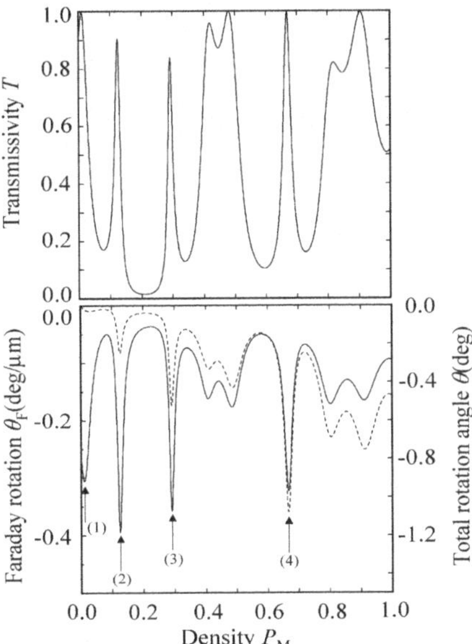

Fig. 2. Transmissivity T, Faraday rotation angle θ_F (solid curve), and total rotation angle $\theta = \theta_F N_M d_M$ (broken curve) versus Bi:YIG density $P_M = N_M d_M / D$ for the film with $N = 16$, $b^{16} = (\text{ABAA})_{\text{hex}}$ and $D = 5$ μm, where λ (wavelength of light) is 1.15 μm.

interference effect. The enhancement in θ_F due to the localization of light can be explained from the fact that the localization conditions for two eigenmodes of lights in Bi:YIG layers (right- and left-hand circularly polarized lights) are slightly different to each other [4].

These results indicate that the stronger the localization of light, the larger the enhancement in θ_F. This is indeed demonstrated in films with the structure of 1-D photonic crystal with a magnetic defect (1-D magnetophotonic crystal) which supports the strong localization of light. Figure 4 shows a typical example of the wavelength spectra of T and θ_F of the 1-D magnetophotonic crystal: The film structure is expresses by $(\text{Bi:YIG}/\text{SiO}_2)^8/\text{Bi:YIG}^2/(\text{SiO}_2/\text{Bi:YIG})^8$ and is designed under the condition of $n_M d_M = n_S d_S = \lambda/4$ ($\lambda = 1.15$ μm) with refraction indices n_M and n_S of the Bi:YIG and SiO_2 layers, respectively. The film exhibits a sharp transmissivity at $\lambda = 1.15$ μm with a huge Faraday rotation: as clearly seen in Fig. 5, θ_F at $\lambda = 1.15$ μm reaches more than -16 deg/μm without causing elliptical polarization of light ($\eta_F = 0.0$ at $\lambda = 1.15$ μm). The localized state of light at $\lambda = 1.15$ μm is illustrated in Fig. 6. In comparison with the case of weak localization of light in Figs. 3(a) and 3(b), light is, in fact, strongly localized within the film, yielding the huge enhancement for the Faraday effect.

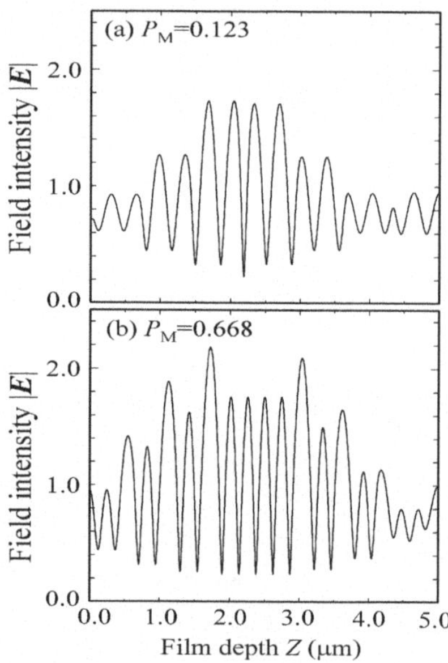

Fig. 3. Distributions of light field $|E|$ in films in Fig. 2 when $P_M = 0.123$ (a) and $P_M = 0.668$ (b). The light submerges into the film at $Z = 0$ and outgoes from the film at $Z = 5$ μm.

3 1-D Photonic Crystal with a Magnetic Defect

In II, the large enhancement in qF due to the localization of light was demonstrated for the film with $(Bi:YIG/SiO_2)^8/Bi:YIG^2/(SiO_2/Bi:YIG)^8$ multilayer structure. Practically, however, it is difficult to form such a film by maintaining the multilayer structure without loosing the good magneto-optical properties of Bi:YIG layers. Favorably, the large magneto-optical effect is also available with a dielectric multilayer film containing only one thin Bi:YIG film as a magnetic defect: Figure 7 shows such a 1-D magnetophotonic crystal, where a thin Bi:YIG layer (thickness d_M) is introduced as a magnetic defect in the SiO_2/SiN periodic multilayer film. In this case, thicknesses of the SiO_2 and SiN layers are designed as $d_{SiO_2} = 108$ nm and $d_{SiN} = 77$ nm, respectively, so as to set the localization wavelength at 650 nm.

The photonic band structure of the medium with $(SiO_2/SiN)^{10}/Bi:YIG/(SiN/SiO_2)^{10}$ structure is depicted in Fig. 8, where the central Bi:YIG layer had the thickness of $d_M = 160$ nm. In this case, the localization wavelength was designated to be 650 nm, so that the medium exhibits the photonic bandgap between 580 nm and 720 nm and the resonant transmission due to the light localization appears at 650 nm. This resonant transmission of light corresponds to the localized mode of light, whose wavelength (650 nm) shifts from shorter to longer wavelengths of light within the photonic bandgap asso-

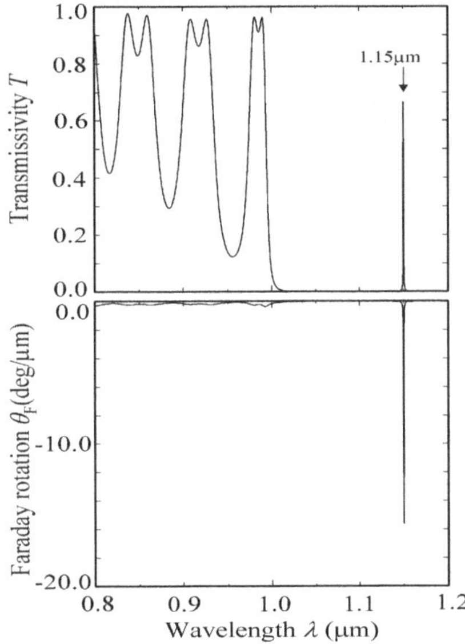

Fig. 4. Wavelength spectra of transmissivity T (above) and Faraday rotation angle θ_F (below) of the multilayer film with $N = 34$ and (B:YIG/SiO2)^8/ Bi:YIG^2/(SiO_2/B:YIG)^8 structure.

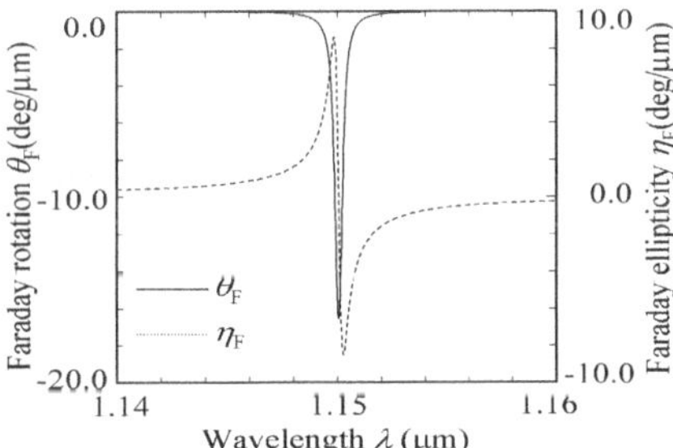

Fig. 5. Faraday spectra (θ_F and η_F vs. λ) near Fabry-Perot resonance wavelength $\lambda = 1.15$ μm.

ciated with increasing d_M. It should be noted that the localized mode shows very large Faraday rotation reaching approximately −8 degrees, and hence in the 1-D magnetophotonic crystal, high transmissivity and large Faraday rotation are fulfilled simultaneously by the use of localized mode. This unique property is indeed attractive for various opto-electronic applications including the optical communication devices and the magneto-optical recording.

In such engineering applications, the most important performance parameter of media is the figure-of-merit parameter Q_F defined by $Q_F = T^{1/2}|\theta_F|$

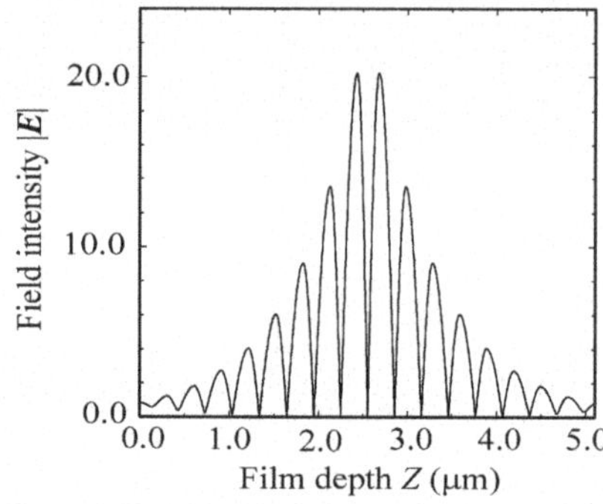

Fig. 6. Field distribution of light at $\lambda = 1.15$ μm.

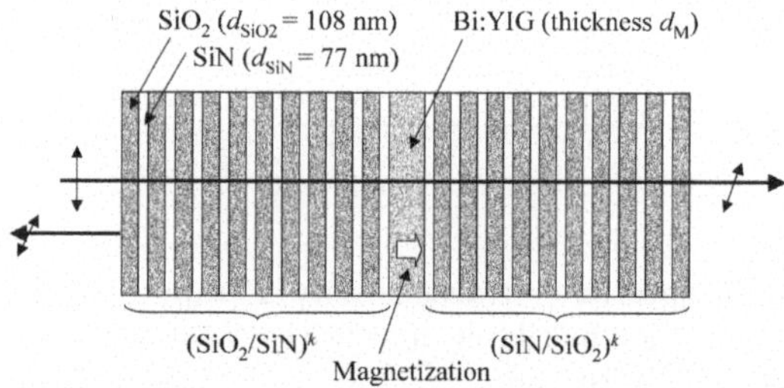

Fig. 7. One-dimensional magnetophotonic crystal, where a thin Bi:YIG film is sandwiched between two dielectric multlayer films.

deg. It is then interesting to evaluate the figure-of-merit parameter of the localized mode. For the 1-D magnetophotonic crystal with $(SiO_2/SiN)^{10}$/ Bi:YIG/$(SiN/SiO_2)^{10}$ structure, changes in Q_F of the localized modes were calculated as a function of the central Bi:YIG layer thickness d_M. The results are summarized in Fig. 9 for three cases of the localized modes with different orders that are distinguished by the labels (0), (1) and (2). In the photonic band gap, those localized modes with different orders appear as follows: When $d_M = 0$ (absence of the Bi:YIG layer), the lowest localized mode (0) exists at the center wavelength of light within the photonic band gap, and shifts toward the lower photonic band edge (longer wavelength of light) with increasing d_M. Once when the lowest localized mode disappears in the lower photonic band, the first localized mode (1) appears from the upper photonic band, and shifts toward the lower photonic band edge with d_M in the same manner as the mode (0).

As seen in the figure, when $d_M \approx 150$ nm, the first localized mode existing at the central wavelength of the photonic band gap shows the transmissivity $T \approx 50$ % and Faraday rotation $\vartheta_F \approx -6$ deg, yielding the figure-of-merit parameter of $\vartheta_F \approx 4$ deg. Generally, Bi:YIG single crystal with approximately

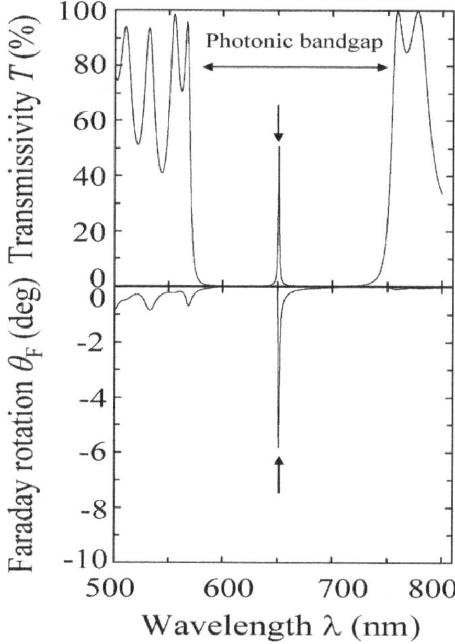

Fig. 8. Photonic band structure of the 1-D magnetophotonic crystal in Fig. 7, where the central Bi:YIG layer had the thickness of 160 nm.

25 μm thickness shows the maximum figure-of-merit parameter of $\vartheta_F \approx 6$ deg. at $\lambda = 650$ nm. Then, almost equivalent value of Q_F is available with the 1-D magnetphotonic crystal whose Bi:YIG layer thickness is less than 1 % of the Bi:YIG single crystal.

4 Formation of Bi:YIG Films

To confirm the theoretical results experimentally, 1-D magnetophotonic crystals with Bi:YIG films were formed by using a RF magnetron sputtering apparatus. To materialize the good optical and magneto-optical properties as the theory predicted, the key is the formation of Bi:YIG films without destroying the multilayer structure. Then, prior to fabricating the 1-D magnetophotonic crystals, preparation of Bi:YIG films was examined. The Bi:YIG films were deposited under the sputtering conditions listed in Table 1: Use of pure Ar as the sputtering gas was very effective to obtain the films with smooth surfaces whose average roughness was within ±2.5 nm. To control the film composition, we used the sputtering target with $Bi_{1.0}Y_{2.5}Fe_5O_x$ composition that is intentionally sifted from the stoichometry composition of garnet. From EDX and magneto-optical measurements (see Fig. 11), the film composition was identified as $Bi_{0.7}Y_{2.3}Fe_5O_{12}$.

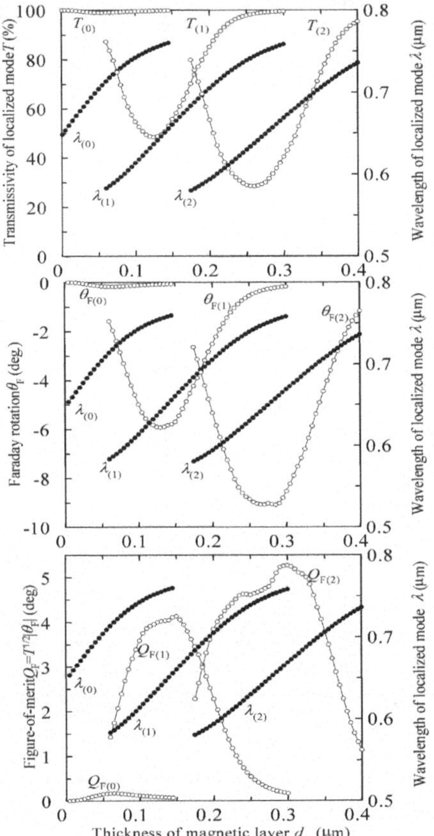

Fig. 9. Transmissivity T, Faraday rotation θ_F and the resultant figure-of-merit parameter $Q_F = T^{1/2}|\theta_F|$ of the localized modes (0), (1) and (2) in the 1-D magnetophotonic crystal with $(SiO_2/SiN)^{10}$/Bi:YIG/$(SiN/SiO_2)^{10}$ structure. In the figure, $\lambda_{(0)}$, $\lambda_{(1)}$ and $\lambda_{(2)}$ are wavelengths of the localized modes (0), (1) and (2), respectively.

Figure 10 is the XRD (Cu-Kα line) patterns of film, where the change in crystallographic structure of films associated with thermal annealing in an electric furnace at 700°C in air is clearly seen. The as-deposited sample had amorphous structure, but was crystallized to single-phase Bi:YIG after the annealing for more than 10 mins. Corresponding to the change in crystallographic structure from amorphous to single-phase Bi:YIG, the annealed samples exhibited good magneto-optical Faraday effect: A typical wavelength spectrum of Faraday rotation of the Bi:YIG film annealed for 10 mins is shown in Fig. 11. The Faraday rotation angle reached approximately 20 deg/μm at

Table 1. Sputtering conditions of Bi:YIG film

Background pressure	3.6×10^{-7} Torr
Sputtering gas	Ar
Sputtering puressure	7×10^{-3} Torr
Target	$Bi_{1.0}Y_{2.5}Fe_5O_X$
Substrate	#7059 glass plate
Substrate temperature	Room temperature
Sputtering power	150 W

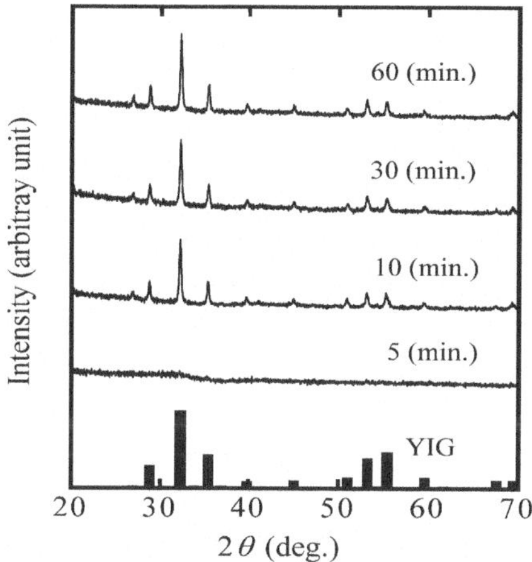

Fig. 10. XRD patterns of Bi:YIG film, where the annealing time was taken as a parameter.

$\lambda = 400$ nm, the value of which is comparable to that of Bi:YIG single crystal.

In the above procedure, we employed the post annealing at 700°C for more than 10 mins. so as to obtain Bi:YIG single-layer films with smooth surface and good magneto-optical properties. However, these annealing conditions are not practically applicable for forming the 1-D magnetophotonic crystals with a Bi:YIG film which is sandwiched between two dielectric multilayer films such as SiO_2/Ta_2O_5. This is simply because such a dielectric multilayer film is unable to stand up to the annealing at 700°C for more than 10 mins.

Then, to solve the above problem, we developed a pulsed-light annealing technique with infrared radiation. Figure 12 shows the schematic drawing of the annealing system: A film sample was placed on a water-cooled sample stand with a thin In sheet. A glassy carbon plate placed on the film surface

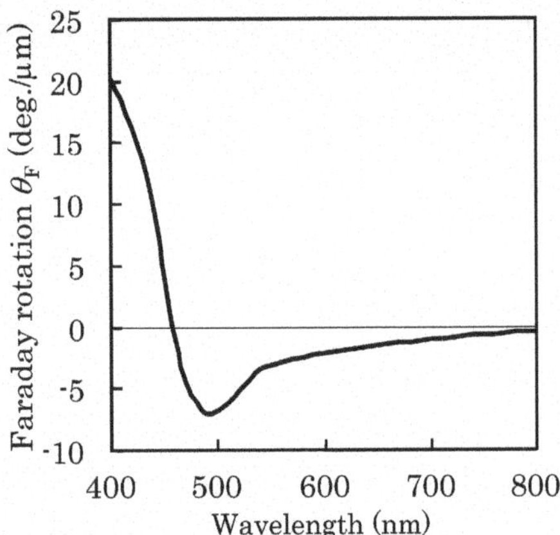

Fig. 11. Wavelength spectrum of Faraday rotation θ_F of the Bi:YIG single layer film annealed for 10 mins.

was heated with pulsed infrared radiation. A typical annealing cycle monitored with a thermocouple detecting the surface temperature of the glassy carbon is depicted in Fig. 13. As seen in the figure, the system enables us to anneal a film sample with very high rates for temperature elevation (40°C/s) and descent (30°C/s).

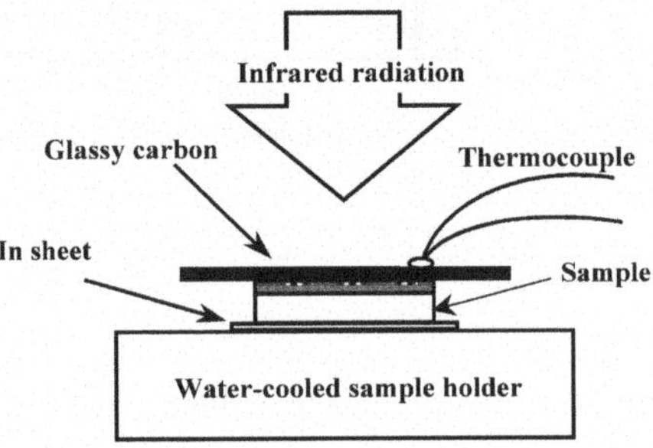

Fig. 12. A schematic drawing of the pulsed-light annealing system.

In Fig. 14, XRD patterns of the post-annealed Bi:YIG films are shown, where the annealing was achieved with the pulsed-light annealing system in Fig. 12. In the figure, annealing time that was measured from the room temperature (see Fig. 13) is taken as a parameter. In this case, films with the single garnet phase were favorably obtained within the very short post annealing of 3 mins. In comparison with the conventional annealing with an electric furnace (Fig. 10), the annealing time is approximately less than 30 %. Corresponding to the change in crystallographic structures of films,

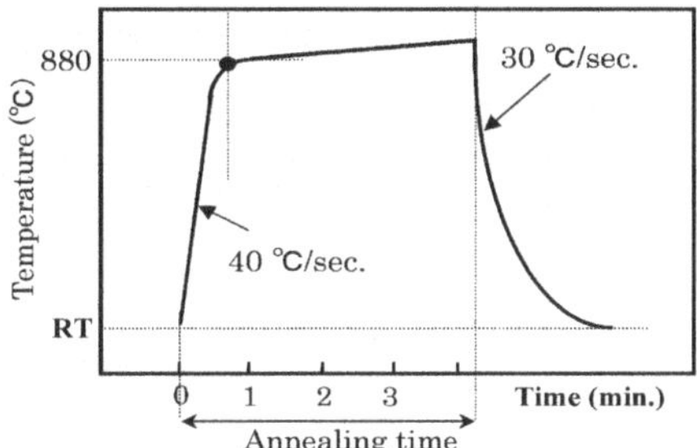

Fig. 13. Typical example of the annealing cycle with the pulsed-light annealing system in Fig. 12. Note that the annealing time in this case was measured from the room temperature (RT).

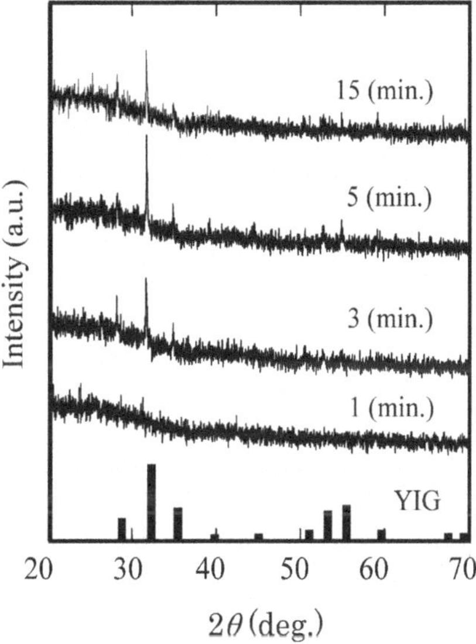

Fig. 14. XRD patterns of Bi:YIG film after post annealings with the pulsed-light annealing technique.

the Bi:YIG films thus annealed were actually exhibited considerably good magneto-optical properties as shown in Fig. 15.

5 1-D Magnetophotonic Crystals

In the 1-D magnetophotonic crystal, we employed the dielectric multiplayer film composed of SiO_2/Ta_2O_5 pairs. To examine the performance of the dielectric multiplayer film and to evaluate the usefulness of the pulsed-light an-

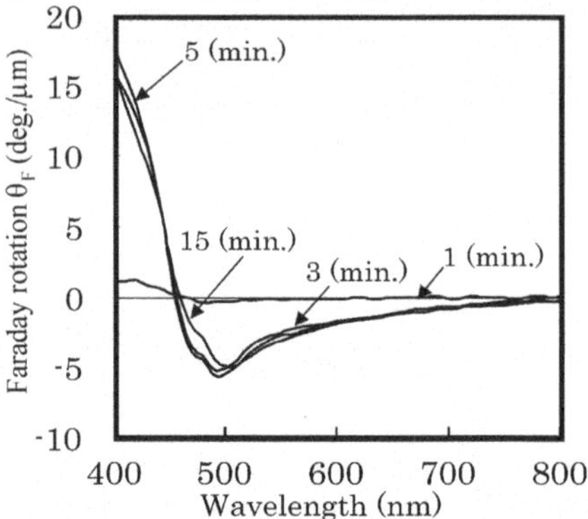

Fig. 15. Wavelength spectra of Faraday rotation for the Bi:YIG films corresponding to the XRD patterns shown in Fig. 14.

nealing technique, 1-D photonic crystals with $(SiO_2/Ta_2O_5)^5(Ta_2O_5/SiO_2)^5$ multilayer structure were first prepared using RF magnetron sputtering with the conditions listed in Table 2. Total thickness of the $(SiO_2/Ta_2O_5)^5$ film deviated approximately 10 % in maximum from the expected value which was determined from the sputtering rates. This caused the slight shift in the wavelength of localized mode: The designated wavelength was 650 nm from the refractive indices of 1.5 for SiO_2 and 1.8 for Ta_2O_5, while the observed wavelength was approximately 780 nm. This situation is clearly seen in Fig. 16, where the wavelength spectra of transmissivity T are shown for the 1-D photonic crystal before annealing (a) and after pulsed-light annealing at 830°C for 15 mins (b). Although the film had the deviation of 10 % in its thickness, the medium exhibited a clear photonic bandgap between 650 nm and 900 nm. The localized modeappeared at approximately 780 nm with the transmissivity reaching 75 %. In comparison with Fig. 16(a) and 16(b), no significant difference in the photonic band structure is seen, suggesting that the multilayer structure was maintained after the pulsed-light annealing.

Based on the above results, 1-D magnetophotonic crystals with $(SiO_2/Ta_2O_5)^5$/B:YIG/$(Ta_2O_5/SiO_2)^5$ structure was formed. The preparation procedure is summarized as follows: The lower dielectric multilayer film, $(SiO_2/Ta_2O_5)^5$, was deposited onto a #7059 glass substrate using RF magnetron sputtering with the conditions listed in Table 2. Then, Bi:YIG film with 167 nm thick was deposited directly onto the dielectric film with the conditions in Table 1. Then, the multilayer film with $(SiO_2/Ta_2O_5)^5$/Bi:YIG structure was annealed with the pulsed-light annealing technique, by introducing the infrared radiation onto the glassy carbon plate placed on the front surface of the Bi:YIG film. The annealing was achieved at 850°C for 15 mins in air. Finally, the annealed film sample was covered with the top multilayer film, $(Ta_2O_5/SiO_2)^5$, by sputtering.

Table 2. Sputtering conditions for SiO_2 and Ta_2O_5 films

	SiO_2	Ta_2O_5
Background pressure	$< 1.0 \times 10^{-7}$ Torr	
Sputtering gas	Ar : O_2 = 9.8 : 0.2	
Sputtering puressure	10×10^{-3} Torr	
Substrate temperature	100°C	
Sputtering power	150 W	
Target	SiO_2	Ta_2O_5
Sputtering rate	2.3 nm/s	2.1 nm/s
Thickness	111 nm	92 nm

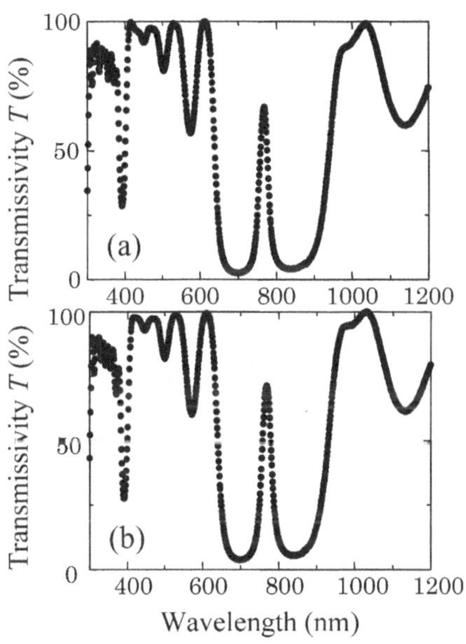

Fig. 16. Wavelength spectra of transmissivity T for the 1-D photonic crystal with $(SiO_2/Ta_2O_5)^5/\ (Ta_2O_5/SiO_2)^5$ structure: (a) before annealing and (b) after pulsed-light annealing at 830°C for 15 mins.

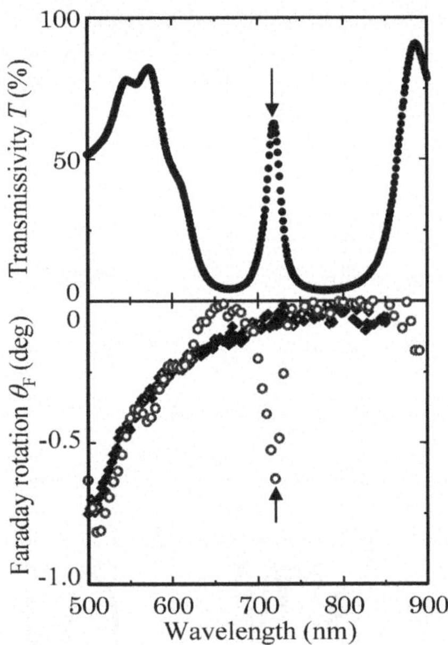

Fig. 17. Wavelength spectra of transmissivity T (above) and Faraday rotation θ_F (below) of the 1-D magnetophotonic crystal with $(SiO_2/Ta_2O_5)^5/$ Bi:YIG$/(Ta_2O_5/SiO_2)^5$ structure.

In Fig. 17, wavelength spectra of transmissivity T (above) and Faraday rotation θ_F (below, open marks) of the medium thus obtained are depicted. For comparison, θ_F-spectrum of the Bi:YIG signle-layer film with approximately the same thickness as that in the 1-D magnetophotonic crystal is also plotted by solid marks. As seen in the figure, the medium exhibited a clear photonic bandgap between 650 nm and 850 nm accompanied by a localized mode at about 720 nm. The photonic band structure was slightly modified form that in Fig. 16 due to the insertion of Bi:YIG layer as a magnetic defect into the dielectric periodic structure. The localized mode showed considerably high transmissivity of $T = 63$ % and large Faraday rotation of $\Theta_F = -0.7$ deg, the rotation angle of which is approximately ten times larger than that of the Bi:YIG single-layer film.

6 Concluding Remarks

Unique optical and magneto-optical properties of 1-D magnetophotonic crystals with Bi:YIG films were studied theoretically and experimentally. Particular features of the 1-D magnetophotonic crystals are summarized as the following: (1) Although only one very thin Bi:YIG film is used in the 1-D magnetophotonic crystal, its localized modes exhibit high transmissivity and large Faraday rotation simultaneously, yielding a high figure-of-merit parameter equivalent to the 100 times thick Bi:YIG single crystal. (2) The wavelength at which the localized mode appears can be modified by merely adjusting the layer thicknesses. In other words, the media provide artificially controllable wavelength spectra of transmissivity and Faraday rotation, which is very much attractive for various opto-electronic applications.

According to recent theoretical calculations by Sakaguchi et al. [14], the 1-D magnetophotonic crystal could possess 45-degree Faraday rotation with the transmissivity higher than 90 %. This performance is indeed satisfactory for realizing low-loss isolators. To obtain such 1-D magnetophotonic crystal, we are now trying to form the media by using a RF dual ion-beam apparatus (TDY: Tokyo Denshi Yakin, Japan). In addition, further interests exist in the magnetophotonic crystals with 2- and 3-dimensional structures, even though these media still have difficulties in their actual formation. To overcome such problems and examine their potentials as a new class of magneto-optical media for next generation, further investigations are now under way.

References

1. *c.f.*, J. Joannopoulos, R. Meade, and J. Winn: *Photonic Crystals*, Princeton University Press, Princeton, NJ (1995).
2. E. Yablonovitch: Phys. Rev. Lett. **58**, 2059 (1987)
3. E. Yablonovitch, T. J. Gmitter, K. M Leung: Phys. Rev. Lett. **67**, 2295 (1991)
4. E. Yablonovitch: J. Opt. Soc. Am. **10**, 282 (1993)
5. H. Nishizawa, T. Nakayama: J. Phys. Soc. Jpn. **66**, 613 (1997).
6. M. Inoue, T. Fujii: J. Appl. Phys. **81**, 5659 (1997).
7. M. Inoue, T. Fujii: J. Magn. Soc. Jpn.**21**, 187 (1997), in Japanese
8. M. Inoue, T. Fujii, K. I. Arai, M. Abe: J. Magn. Soc. Jpn. **22**, 141 (1998)
9. M. Inoue, K. I. Arai, T. Fujii, M. Abe: J. Magn. Soc. Jpn. **22**, 321 (1998), in Japanese
10. M. Inoue, K. I. Arai, T. Fujii, M. Abe: J. Appl. Phys. **83**, 6788 (1998)
11. M. Inoue: J. Magn. Soc. Jpn. **22**, 1105 (1998), in Japanese.
12. M. Inoue, K. Matsumoto, K. I. Arai, T. Fujii, M. Abe: J. Magn. Magn. Mat. **196–197**, 611 (1999)
13. M. Inoue, K. I. Arai, T. Fujii, M. Abe: J. Appl. Phys. **85**, 5768 (1999)
14. S. Sakaguchi, N. Sugimoto: Optics Communications **162**, 64 (1999)

Part II

Nano-Patterned Media

Selective Removal of Atoms as Basis for Ultra-High Density Nano-Patterned Magnetic and Other Media

Boris Gurovich, Evgenia Kuleshova, Dmitry Dolgy, Kirill Prikhodko, Alexander Domantovsky, Konstantin Maslakov, Evgeny Meilikhov, and Andrey Yakubovsky

Russian Research Center "Kurchatov Institute",
Kurchatov sq. 1, Moscow 123182, Russia

Summary. The paper demonstrates a possibility for effective modification of the thin-film material' chemical composition, structure and physical properties as result of selective removal of atoms by the certain energy ion beam. One of the most promising results of this effect is a production of devices with nanostructured high-density patterned magnetic media.

1 Introduction

The crossover at the high-technology market from the current micro-technologies to the future nano-technologies is expected in the nearest 5–10 years. This process, revolutionary in its scope and consequences, will allow to derive all the benefits from both scaling and new circuit technology, based on the new physical principles, which can be materialized only in nanoscale devices. The most promosing way for ultra-hight-density devices prodoction is the patterned media developing. This commercial task depends directly on the research advances.

We have discovered and studied in detail the new fundamental effect of "selective removal of atoms by ion beams", which lays the foundation of the new technology for micro- and nano-devices production [1, 2].

Traditionally, materials with different physical properties are those with various chemical compositions. We have experimentally demonstrated the possibility of direct alteration of a solid atomic composition under exposure to an accelerated ion beam of a specific energy. This modification of atomic composition is not a result of any chemical or nuclear reaction, but is completely caused by the selective removal of atoms of a specific type from two- or multi-atomic compounds as a result of atomic displacements induced by the accelerated ions. Such modification can result in a radical change of the material' physical properties and, in particular, to the transformation of insulators into metals or semiconductors, nonmagnetic materials into magnetic ones, changing optical properties, etc. [2]. It means that there is an opportunity to produce in thin films (10–1000 nm) a desirable pattern in local areas having various atomic compositions.

The physical basis of the method is as follows. Let us consider a situation that arises during interaction of a monochromatic ion beam of energy E and mass m with a two-atomic crystal consisting of atoms of different masses M_1 and M_2. The maximum energy transferred by the ions to atoms of a crystal is [3]:

$$E_{max}^{(1,2)} = \frac{4mM_{1,2}}{(M_{1,2}+m)^2} \cdot E, \tag{1}$$

where $E_{max}^{(1)}$ and $E_{max}^{(2)}$ are maximum energies which could be transferred by the accelerated ions to atoms with masses M_1 and M_2.

In deciding on a particular material for selective removal of atoms, metal compounds that are insulators in the initial state hold the greatest practical interest. Films behaviour of different insulating di- and polyatomic materials under irradiation depends strongly on the ions energy, all other factors being the same. The observed radiation-induced modification of thin film properties was of clearly defined threshold character [4]. No modifications of structure, composition, electrical and magnetic properties of thin films were detected until the ion energy reached certain minimum value (which is individual for every studied material). It means that as long as the energy E_{max} transferred to material atoms by ions is low ($E_{max} < E_{d1}$ and $E_{max} < E_{d2}$, where E_{d1}, E_{d2} are the displacement threshold energies for atoms of the first and the second types, respectively), atom displacements off regular lattice positions do not take place.

With increasing the ions energy, conditions appear when $E_{d1} < E_{max} < E_{d2}$. Under this condition the selective removal of light atoms of the first kind (e.g. oxygen, nitrogen or hydrogen) is observed from the studied compounds (oxides, nitrides or hydrides, respectively). As a result the dramatic variations of the structure, composition, as well as electric, magnetic and optical properties appear as will be shown below.

With further increasing ion energy, the condition $E_{d1} \leq E_{d2} < E_{max}$ is reached and atoms of both types in diatomic compounds begin to displace off the regular lattice positions (but with various efficiency). The further increase in E_{max} has only small influence on the selectivity of atomic displacement in the studied energy range.

The initial insulating films of different materials were in one of three structural states:

1. Polycrystalline state with grains of 5–100 nm.
2. Amorphous state.
3. Mixed state with randomly oriented grains distributed in amorphous matrix or separated by thin amorphous layers.

Electron diffraction patterns of all the above-mentioned types of films correspond to the annular diffraction specific for polycrystals or look like a diffuse halo or, at last, appear to be a composition of both those types of diffraction.

It should be also noticed that the most of studied films have crystal structures, which do not correspond to the handbook data for the same stable-state compounds. It is known that similar situation is characteristic for thin films and associated with their intrinsic inclination for polymorphism and no equilibrium phase formation [5]; the same could lead to different anomalies in thin film properties.

Experiments show that, independent of the original insulator structure, they behave very similar in the course of selective removal of atoms. As a rule, at the first stage of polycrystal or mixed structure insulator irradiation the amorphization of crystallites takes place. That effect is detected clearly by disappearing crystal contrast in the dark-field electron-microscopic images (see Fig. 1a). With this, significant decrease in intensity and broadening of point reflexes (lines) in annular electron diffraction patterns (till their complete transformation into diffuse halo characteristic to amorphous materials) are observed (see Fig. 1b,c).

Further irradiation is accompanied by the next phase transformation corresponding to the transition from amorphous state to crystal one. This is evident due to appearance of crystal contrast in the dark-field electron-microscopic images associated with newly generated grains (see Fig. 1a). Simultaneously, an absolute or partial (depending on original material) disappearance of diffuse halo, increasing intensity and appearing new system of diffraction rings in electron diffraction patterns occur, that testify appearing a new (metallic) phase (see Fig. 1c,d). The formation of a metallic phase after irradiation with high fluencies is demonstrated below when presenting the obtained results.

In some cases, diffraction in crystallites of newly generated metallic phase does not correspond to those types of crystal lattices, which are cited in handbooks for relevant pure bulk metals. At the same time, diffraction in crystallites of new metallic phase (produced from insulators due to selective removal of atoms) in some instances corresponds to diffraction data, which have been obtained for thin films of relevant pure metals. For example, with selective removal of oxygen atoms out of WO_3, metallic tungsten has been generated in *fcc*-lattice ($a = 4.19 \pm 0.02$ Å) instead of *bcc*-lattice typical for bulk tungsten. As mentioned above, that could be associated with polymorphism characteristic for thin films. However, in some papers (see, for instance [6]) the formation of thin tungsten films with *fcc*-lattice ($a = 4.15$ Å have been observed as a result of pure tungsten sputtering. Taking into account the estimated accuracy of these electron diffraction measurements ($\sim 2\%$), one should consider the coincidence of lattice parameters as very good. Nevertheless, in some cases the metal films produced by selective removal of atoms have got crystal lattice identical to that of the same bulk metal. For example, copper produced out of CuO demonstrates the diffraction characteristic

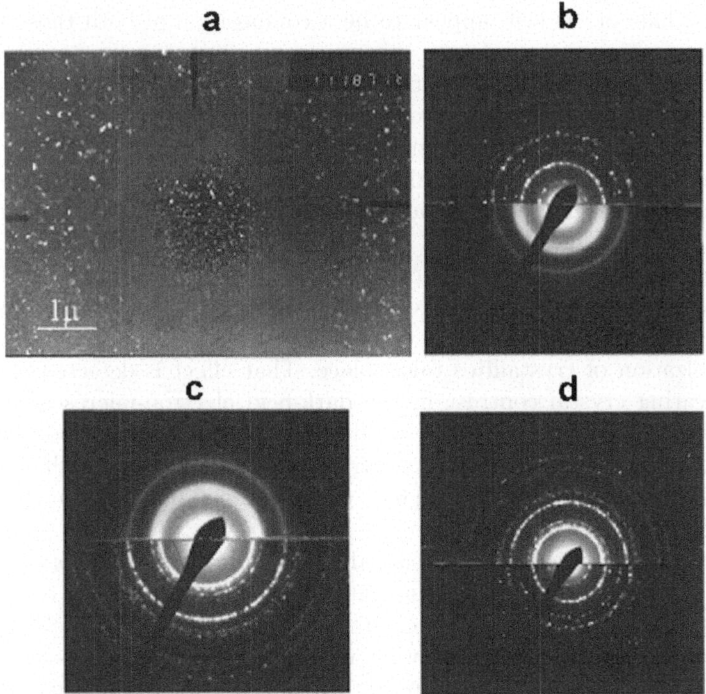

Fig. 1. Structure and diffraction changes in Lanthanum hydride during Selective Removal of Atoms process under 200 keV electron irradiation in electron microscope: **(a)** – dark field image of transformed material (metal) (center), initial material (periphery) and transition amorphous area; **(b)** – diffraction patterns of initial and amorphous material; **(c)** – diffraction patterns of transition amorhous material and metal; **(d)** – diffraction patterns of initial material and metal.

of *fcc*-lattice with parameter $a = 3.60 \pm 0.02$ Å while the handbook value is $a = 3.615$ Å.

Thus, by varying the mass and the energy of ions, it is possible to achieve the situation in a two- or a multi-atomic crystal when the higher energy would be transferred to the atoms of low or high masses. If the maximum transferred energy exceeds the threshold value E_d for atoms of only one kind, then there exists a method of selective removal of only light (or only heavy) atoms from two- or a multi-atomic crystal.As a rule, this energy $E_d \approx 20$–25 eV, which exceeds considerably the sublimation energy [7].

The considered mechanism regarding the displacement of atoms of different kinds in a crystal refers equally to the same compounds in an amorphous state. Thus, it's clear that at the normal incident of the ion beam on a crystal surface, it's possible to achieve conditions when selective removal of only one kind of atoms is observed (for $E_{d1} \leq E_{max} < E_{d2}$). Under the condition $E_{d1} < E_{d2} < E_{max}$ (when the displacement selectivity is provided by the difference in the displacement rates of the various atom kinds) it's possible to remove the selected kind of atoms up to the needed properties level of the material composed of the residuary atoms of the second kind within a layer of a thickness comparable with the ion projective length in a two- or a multi-atomic crystal.

Let us formulate some obvious features of the considered physical mechanism of selective removal of atoms:

- The rate of the process is proportional to the flux density of the incident ion beam.
- The process can be proceed in a layer below the surface even if covered by another material, if its thickness is less than the ion projective length in the layer. If, in addition, the threshold energy of atomic displacement in the additional layer is higher than the transferred energy from the ions, the directed displacement of atoms in that material will not occur. Otherwise, the atoms of material penetrate in the underlying layer and their transfer in the beam direction occurs over a distance comparable with the ion projective length in the "sandwich" considered.

The above-mentioned features of the proposed method determine its potential for efficient, purposeful, and spatially modulated modification of the composition, structure, physical, and chemical properties of materials. It will be shown below that such a modification of chemical composition can dramatically change the physical properties of a thin material layer, e.g. to produce an insulator-metal transition, to change magnetic or optical properties and so on. Thus a possibility exists to create a controlled volume "pattern" of areas with different physical properties, in particular for production of single-domain patterined magnetic media with high areal density.

2 Experimental Results and Discussion

Among the diatomic compounds, those are, for example, many metal oxides, as well as some hydrides and nitrides of metals. Though in the course of this work experiments were performed with compounds of all above-mentioned types, metal oxides were investigated most thoroughly. Qualitative features of the effects accompanying selective removal of atoms are identical in compounds of all above-mentioned types. The aim of experiments was to remove selectively oxygen (nitrogen or hydrogen) atoms by irradiation of the original insulator and to obtain finally a metal. Experiments have been performed with thin films of different thickness, which have been produced by reactive metals sputtering in the atmosphere of relevant gases (oxygen, nitrogen or hydrogen) [5].

In most cases, films have been irradiated by protons with energies $\sim$ 1–5 keV. In addition, in some experiments films have been irradiated in electron microscope column by electrons with energy of 100–200 keV.

During our study the complex investigations of initial and irradiated films were performed which included: measurement of electric resistivity within the temperature range 4.2–300 K, magnetic and optical properties [5], structural measurements by means of transmission electron microscopy and electron diffraction analysis [8, 9], as well as by methods of X-ray photoelectron spectroscopy [10].

To measure the film electrical resistance in the course of the ion irradiation, special through electrical contacts were prepared in the substrates. One of the contact ends was polished abreast with the surface of the substrate. Films were deposited on different insulating substrates with high resistivity ($> 10^{11}$ Ohm $\cdot$ cm), for instance, on diamond-like coat or glass. Reference experiments with irradiating pure substrates (without films) show that radiation-induced effects on their resistance were negligible as compared to those of studied films.

During the process of selective removal of atoms from various compounds, the material volume can change (as compared to the original volume) due to a bulk relaxation of metal atoms into voids created by removal of other atoms. The measurements show that after irradiation the thickness reduction of an irradiated section is about 40–60% (depending on the chemical composition of the compound). It is essential that in spite of such a significant film thickness reduction, there are no changes of linear film dimensions in the plane. The latter is in agreement with the fact that in our numerous experiments we observed neither film exfoliation nor violation of its continuity. It is caused by radiation creep of the irradiated film [11].

In Table 1 are shown results of resistivity measurements for metal films produced by ion sputtering of pure metal targets and for the same metals obtained by selective removal of oxygen atoms out of oxides by proton irradiation. For comparison, standard resistivity values for bulk metals are also presented. The data in Table 1 show that resistivities of thin metal films are strongly dependent of their thickness. As this takes place, the smaller film thickness, the greater deviation from standard values for bulk samples. For films of 10 nm thickness, those distinctions are on order of value, and even more for films of 5 nm thickness. It is, probably, associated with the classic size effect [12]. There is no escape from the notice that in metal films generated by selective removal of oxygen atoms out of oxides (in the same thicknesses range 10–100 nm) resistivity practically does not dependent on the thickness (see Table 1). As has been noticed, it is associated with a small value of electron free path as compared with the film thickness.

Electron-microscopic investigations demonstrate that significant distinctions of resistivity for pure metal films of minimum thickness are most likely connected with the amorphous component in the film structure. In such films

Table 1. Specific resistivity of metal films (μOhm · cm) produced by various methods as a function of their thickness ($T_{meas.}$ = 20°C).

Material	Film thickness, nm				
	5	10	20	50	100
Cu[a]	44	14	9.3	7.4	6.2
Cu[b]	–	27	19	–	–
Cu[c]	1.68				
W[a]	163	133	–	105	65
W[b]	–	800	–	960	700
W[c]	5.39				
Co[b]	–	120	–	–	–
Co[c]	6.24				
Fe[a]	75	35	26	–	–
Fe[b]	–	368	–	–	–
Fe[c]	9.72				
Al[a]	134	16.9	–	11.1	10.1
Al[c]	2.73				

[a] – metal obtained by reactive ion sputtering of pure metal
[b] – metal produced by selective removal of oxygen atoms out of oxide
[c] – standard value for bulk metal

(especially, in the thinnest ones) the volumes of amorphous and crystalline phases are comparable as follows from corresponding electron diffraction patterns. With increasing thickness of pure metal films, the volume part of amorphous phase diminishes sharply and for thickness of 50–100 nm it disappears completely. Simultaneously, with lowering amorphous phase, enlarging of the mean crystallite size in pure metal films is observed.

Effects of thickness influence on the pure metal film structure are most likely conditioned by difficulty of grains' growth in the course of deposition due to surface proximity that promotes conservation of amorphous phase in the films. Besides, it is known that crystallite growth in thin films depends strongly on conditions of their condensation (firstly, on the temperature and the deposition rate) [5, 13].

There is a different situation if one obtains metal films by selective removal of atoms out of insulators – grains' growth is conditioned by crystallite generation from the amorphous state during phase transition. Our experiments show that the mean crystalline size and the existence (or absence) of amorphous component do not depend on the thickness of irradiated insulator films. There is a clear tendency: the lower is melting temperature of the metal produced by insulator film irradiation, the larger is the crystalline size (and, respectively, the smaller is the part of amorphous phase) and the lower is the difference between measured resistivity and the standard one (see Table 1). The effect of melting temperature on crystalline size and disappearance of amorphous phase is mostly evident for copper produced by selective removal of oxygen atoms out of oxide. Copper has the lowest melting temperature among all the studied metals. That is, likely, the reason for the similarity of all its properties to that of the films produced by ion sputtering of pure copper.

It seems interesting to investigate the influence of proton flux on the rate of resistance variation during selective removal of atoms out of oxides and their transformation into metal. Experiments demonstrate that the rate of insulator transformation into metal is proportional to the proton flux. That effect is readily seen under analysis of dose dependencies of resistance [2].

One could also expect that with selective removal of oxygen atoms out of initially nonmagnetic (or weak magnetic) oxides of ferromagnetic metals, films could transform into magnetic state. It is clear that in relevant experiments other di- or polyatomic compounds of ferromagnetic metals could be employed along with oxides.

Below represented are the results of our experiments performed with certain di- and polyatomic systems. Figure 2 illustrates typical changes of a chemical composition during the selective removal of atoms. Figure 3 shows the typical experimental results on both electric (a) and magnetic (b) property modifications of metal oxides during the proton irradiation. As a result, the insulator (MoO_3) transforms into metal (Mo), and non-magnetic material (Co_3O_4) becomes magnetic (Co). Figure 4 demonstrates formation of optical patterned media due to refraction index, as well as reflection and absorption factors variation under the proton irradiation.

It's important to note that selective removal of atoms allows one to simultaneously change the physical properties of separate layers in a multi-layer structure. This is a principal advantage of the proposed technology compared to any other known technology or physical principle. As a result it enables the simultaneous (in parallel) production of structures with different shapes and properties in various layers by ion irradiation through the same mask. Such a procedure allows one to get an overlapping of the structure elements in various layers with an accuracy of about 1 nm. The latter feature is a crucial point in the production of multi-layer nanostructures.

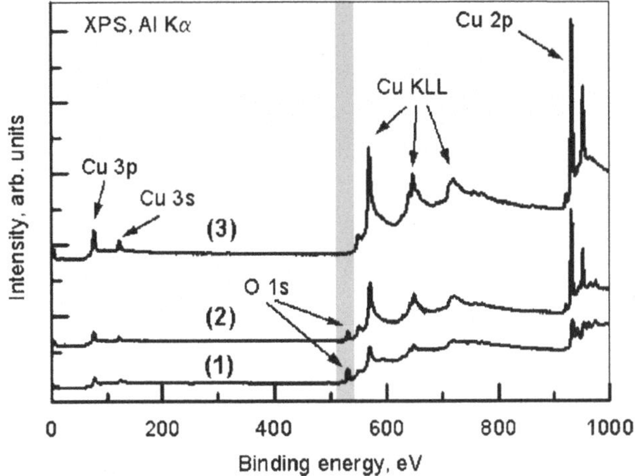

Fig. 2. XPS-spectra of CuO-film in the initial state **(1)** and after proton irradiation of various doses **(2,3)**, which demonstrate the reduction of oxygen concentration in the film during irradiation.

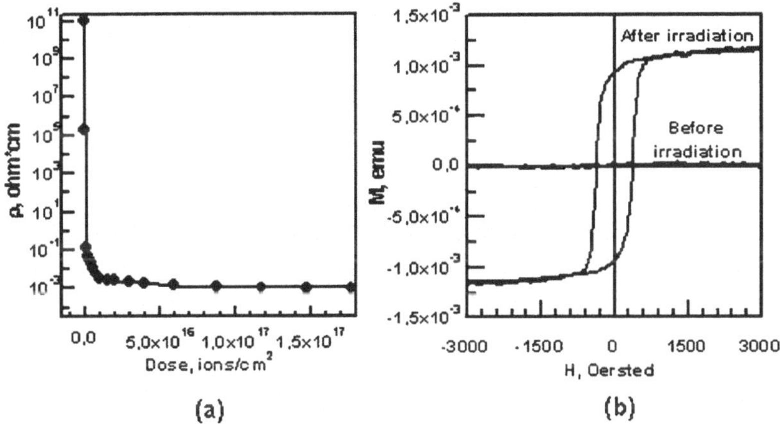

Fig. 3. Typical behaviour of electric resistivity **(a)** and of magnetic hysteresis loop **(b)** of metal oxides during proton irradiation, which demonstrates the transition of insulators into metals **(a)**, and non-magnetic materials into magnetic ones **(b)**.

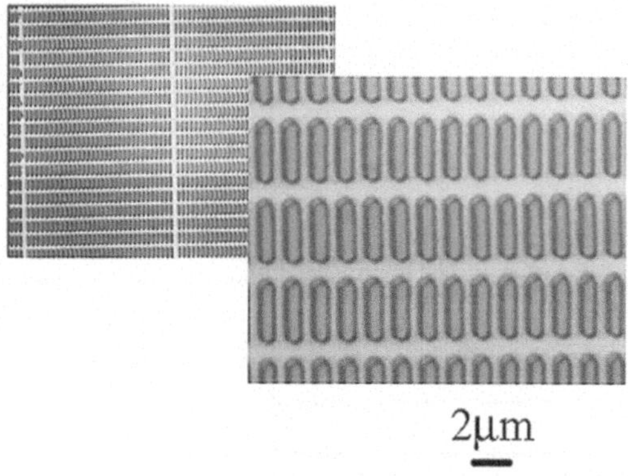

Fig. 4. Optical image of periodic structure formed by Selective Removal of Atoms in tungsten oxide.

In Fig. 5, the experimental results are shown which demonstrate the possibility of simultaneous changes of physical properties in various multi-layered structures with alternating functional and auxiliary layers (the scheme of the sandwich with different layers irradiated through the same mask – a, b). Also shown in Fig. 5 is the dose dependence of Co_3O_4 and CuO layer resistances in the sandwich during their consecutive transformation under irradiation into Co and Cu, respectively (c).

It is very promising to use the proposed method for formation of magnetic patterned media with high areal density. To attain an extreme density of data storage with magnetic recording media, it is necessary to use the patterned media consisting of regularly positioned magnetic nanogranules ("bits") of identical form and orientation. The smallest possible size of the bit is defined by the so-called superparamagnetic limit. The minimum distance between the bits depends on dipole-dipole inter-bit interaction and cannot be less than its lesser size.

On the other hand, one should not arrange the bits so closely that they nearly touch each other due to the above-mentioned magnetic inter-bit interaction. Thus, two physical reasons confine the accessible bit density: superparamagnetic limit and inter-bit interaction. There is the optimal inter-bit distance (and, accordingly, the areal density of the data storage). According to calculations, the optimal gap between the nearest rectangular bits with the shape anisotropy factor of 5–6 amounts about the width of the bit. The calculation shows that with Fe-bits of 3×4 nm^2 the storage density of about 3000 Gbit/inch^2 could be attained.

Elongated (anisotropic) single-domain Co bits of small size (from 320 (1600 nm down to 15 (45 nm) in Co_3O_4 matrix have been produced through a mask prepared by electron lithography. Examples of the structures are

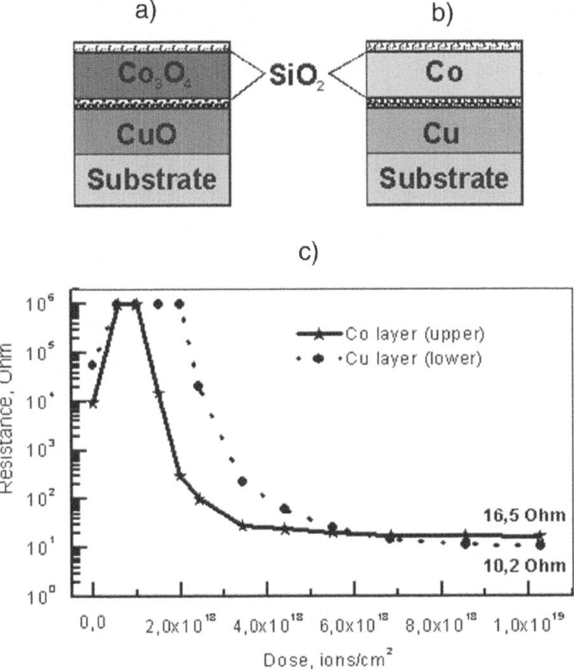

Fig. 5. Schemes of multilayer structure in the initial state **(a)**, after proton irradiation **(b)** and the dose dependence of Co_3O_4 and CuO layer resistances in the sandwich during their consecutive transformation under irradiation into Co and Cu, respectively **(c)**.

shown in Fig. 6,7. These images were obtained by atomic and magnetic force microscopy.

The ranges of bit size and geometries have been studied for patterned structures for which their single domain nature is kept. Investigation of the samples with micron size bits obtained by selective removal of atoms was performed just after the irradiation and demonstrated that these bits are of multi-domain nature which agreed with the published data for maximum size of single-domain particles. Figure 8 shows the topographic and MFM images of magnetic bits of $2 \times 12 \times 0.025$ μm^3 just after the irradiation. Their multi-domain origin is well seen in demagnetized state.

Then those bits were studied after magnetization in external magnetic field directed along and perpendicular to their large side (Fig. 9). The corresponding MFM image demonstrates a pseudo-single-domain origin, i.e. every bit has two well defined poles with no visible domain structure inside the.

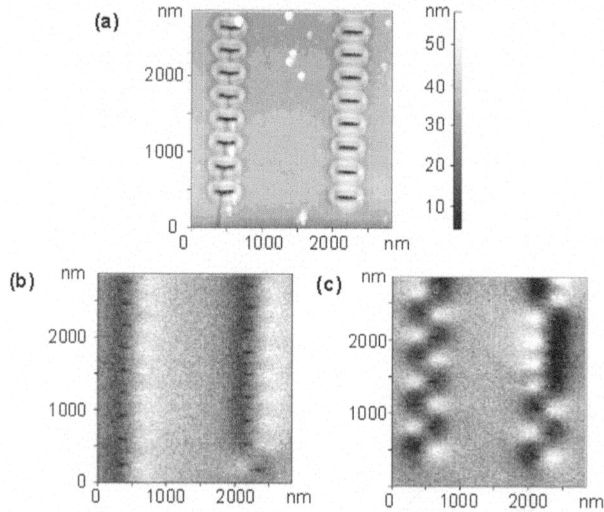

Fig. 6. Examples of patterned magnetic media with areal densitiy of 1.2 Gb/inch^2 (bit size - 80×400 nm^2): **(a)** – AFM topography image, **(b,c)** – magnetic force microscopy images with the different magnetization direction in single domain bits.

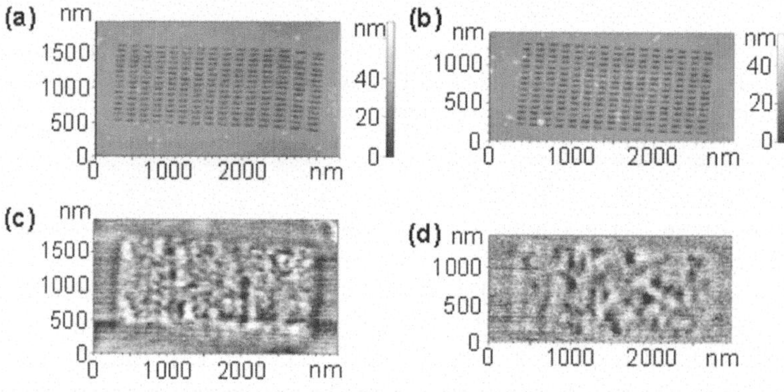

Fig. 7. Examples of patterned magnetic media with areal densities: 45 Gb/inch^2 (bit size - 25×125 nm^2) – **(a,c)** and 57 Gb/inch^2 (bit size - 20×100 nm^2) – **(b,d)**. **(a,b)** – AFM topography images, **(c,d)** – MFM images.

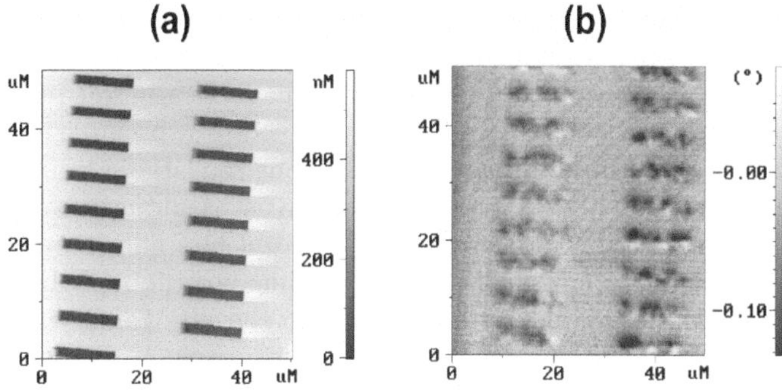

Fig. 8. Topographic (**a**) and MFM (**b**) images of magnetic bits of $2 \times 12 \times 0.025$ μm^3 just after the irradiation (in a virgin state).

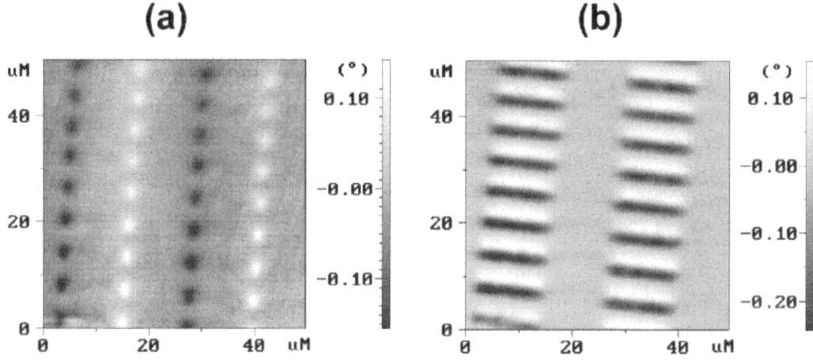

Fig. 9. MFM images of magnetic bits $2 \times 12 \times 0.025$ μm^3 after magnetization along the large (**a**) and short (**b**) side of the bit.

The samples magnetized in this way were kept several days to stabilize their magnetic state. It was demonstrated that such a stabilization didn't influence the MFM image, i.e. the magnetic state was stable.

A similar situation exists for bits $8 \times 48 \times 0.025$ μm^3 where a pseudo-single- domain structure was also observed. But for this sample magnetized in perpendicular direction there is a certain nonuniformity of MFM contrast at magnetic poles. In fact, it probably means a weaker stability of the perpendicular magnetized bit state to decay into multi-domain structure.

Investigation of the bits with larger volume ($10 \times 40 \times 0.045$ μm^3) demonstrated that they have a multi-domain magnetic structure for both lohgitudinal (along the large bit side) and perpendicular (along the short side) magnetizations. However, additional magnetically soft precoat of 50 nm under a structure of the bits $10 \times 40 \times 0.045$ μm^3, leads to pseudo-single-domain structure in the case of longitudinal magnetization while keeps a multi-domain structure at perpendicular magnetization.

The investigation of the bits prepared on magnetically soft precoat in the form of checkerboard structure (Fig. 10) has shown a pseudo-single-domain MFM contrast (Fig. 10b) for magnetization along the large side of the bit. But an evident interaction of adjacent bits was seen as "deformation" of magnetic poles in the direction towards the nearest adjacent bit. This interaction is the strongest one in the point with minimal distance to the next bit, i.e. in its corner. In case of perpendicular magnetization the MFM image has complex multi-domain structure (Fig. 10c).

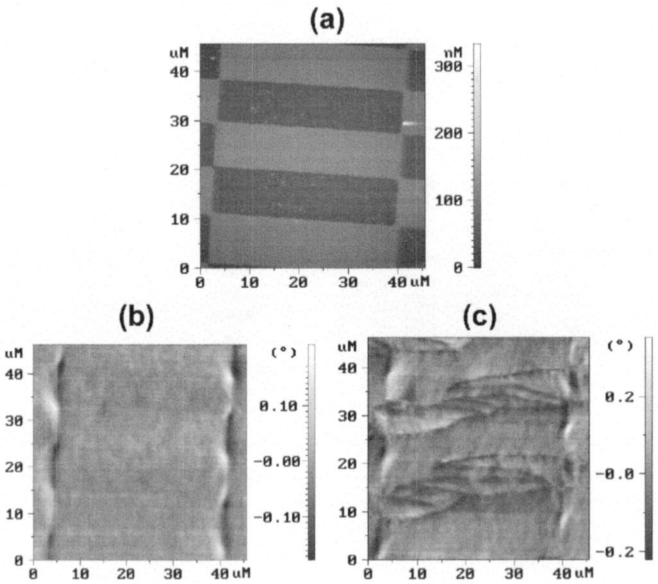

Fig. 10. Topographic **(a)** and MFM **(b,c)** images of magnetic bits $10 \times 40 \times 0.030\ \mu m^3$ prepared as checkerboard structure on the magnetically soft precoat of 50 nm. Magnetization of the bits was along large **(b)** or short **(c)** side of the bit. On topographic image the magnetic bits are located in the dark parts of the image.

It should be noted that the multi-domain image in case of magnetization along the short side of the bit in the checkerboard structure has not random structure inside a separate bit but forms a periodic structure which spreads over the whole sample through the adjacent bits.

The results obtained for checkerboard structures of magnetic bits demonstrate that the bit's magnetic state under magnetization depends on both the parameters of an individual bit (size, anisotropy etc.) and the collective interactions between the bits. It was also shown that the use of magnetically soft precoat allows one to keep a pseudo-single-domain state of the bits stable after their longitudinal magnetization even for large volume of the bits.

3 Perspectives

Nowadays there is only submicron industrial technology for microelectronic device production (with the minimum design rules – 0.065 μm) based on optical lithography. There is the absolute necessity in industrial lithography for producing devices with smaller element size. Many experts consider nanoimprint lithography as the most realistic alternative to optical methods [14]–[18].

Nanoimprint lithography allows one to solve the first problem just now: the single-layered nanostructures with the element size of less than 10 nm [14, 15], and with the areal density of elements $\sim$ 700 Gb/inch^2 [19] have already been produced.

But even nanoimprint lithography itself implies production of multilayer devices in a consecutive manner similar to optical lithography (i.e., consecutive creation of metal, dielectric and other layers and intermediate structures). Extending traditional consecutive principles of microdevice manufacturing to multilayer nanodevices demands the solution of two basic problems: a creation of structural elements with the sizes of 10 nm and less, and ensuring the overlapping of the structure elements in various layers with accuracy of 2–3 nm. Within the frameworks of optical lithography and of traditional consecutive layering principles, the solution of these problems is not obvious now, nor will it be in the foreseeable future.

It can be done by only the combination of nanoimprint lithography and the method of selective removal of atoms. The latter allows one to create various patterns simultaneously in several layers through the same mask. As a result, self-overlapping with $\sim$ 1 nm accuracy can be obtained. Besides this practical reason the application prospectives of this method are conditioned by a number of other reasons.

Ion beams have a few important advantages:

- Negligible back-scattering effects results in increasing spatial resolution of patterns on usual thin films deposited on massive substrates or thin layers on/in massive samples.
- Short wavelengths of incident particles important for high resolution, could be obtained with low accelerating voltages.
- It is important that during irradiation ions leave the material due to diffusion without any negative influence on material properties.

In conclusion, the method could be used to create directly the needed spatial modulations of atomic composition and physical properties of a material, such as metal or semiconductor patterns in insulators, magnetic drawings in nonmagnetic substances, light guides in opaque media, etc. and could be used for fabrication of micro- and nanoscale devices for various applications.

The main advantages of the proposed method are as follows:

- It is a parallel processing technique with respective high throughput.
- Possibility exists to form the various nanopatterns with needed physical properties in different layers through the same mask.
- An intrinsic feature of the method is the self-overlapping ($\sim$ 1 nm) of elements in different layers of the structure.
- The method can be easily combined with traditional CMOS technology to produce hybrid devices (the nanostructures being prepared by the proposed technology.)

We have already successfully tested this method for some elements production of the future nanodevices, the following results being obtained [20, 21, 22]:

- The resolution of 15 nm was achieved, which can be improved up to 3–4 nm in the nearest future.
- The possibility of given relief production on the solid surface with 15 nm resolution has been demonstrated and the prototype of 3.5 inch stamp for imprint lithography was prepared with 1.5 μm resolution.
- The possibility of parallel and simultaneous modification of material' properties in various layers of thin-film multilayer structure was experimentally shown.
- It has been demonstrated that nanopatterns of areas with needed physical properties can be produced in various layers of thin-film multilayer structure through a single mask.
- The possibility of patterned magnetic, conducting and optical medias production was demonstrated with areal density of elements up to 60 Gb/inch^2.

Our future plans address the production of patterned magnetic media with high areal density for MRAM. In this case the magnetic structures will produced in one layer while the corresponding electric elements and connections will be produced in adjacent layer. Our method of selective removal of atoms is the most effective just for this type of multilayer structures with necessity of matching the elements in various layers.

References

1. B.A. Gurovich et al.: Patent US 6,218,278 B1, priority May 1998.
2. B.A.Gurovich, D.I.Dolgy, E.A.Kuleshova, E.P.Velikhov, E.D.Ol'shansky, A.G.Domantovsky, B.A.Aronzon, E.Z.Meilikhov, Physics Uspekhi **44**(1) (2001) 95.
3. L.D.Landau, E.M.Lifshitz, Mechanics. 3rd edition, Vol.**1**, Butterworth-Heinemann, 1976.
4. M.W.Thompson, Defects and Radiation Damage in Metals. Cambridge Press, 1969.

5. Handbook of Thin Film Technology (Eds. L.I.Maissel, R.Gland), New York: McGraw-Hill, 1970.
6. K.L.Chopra, M.R.Randlett, R.N.Duff, Appl.Phys.Lett. **9**, 402, (1966).
7. Handbook of Physical Quantities. (Eds. I.S.Grigoriev, E.Z.Meilikhov), Published by CRC Pr, 1996.
8. P.B.Hirsch, A.Howie, R.B.Nicholson, D.W.Pashley, M.J.Whelan, Electron Microscopy of Thin Crystals, London, Butterworth, 1965.
9. R.Wiesendanger, H-J.Guntherodt (Eds.) Scanning Tunneling Microscopy II: Further Applications and Related Techniques, Springer Series in Surface Science, **28**, Berlin: Springer, 1992.
10. Practical Surface Analysis by Auger and X-ray Photoelectron Spectroscopy (Eds. D.Briggs, M.P.Seah), New York: Willey, 1983.
11. Physical Metallurgy, 3rd edition, (Eds. R.W.Cahn and P.Haasen), North-Holland Physics Publishing, 1983.
12. A.A.Abrikosov, Fundamental of the Theory of Metals, Amsterdam: North-Holland, 1988.
13. K.L.Chopra, M.R.Randlett, R.N.Duff, Philos. Mag., **16**, 261, (1967).
14. S.Chou and P.Krauss, Microelectronic Engineering **35**, 237, (1997).
15. S.Chou, P.Krauss, W.Zhang, L.Guo and L.Zhuang, J. Vac. Sci. Technol. B **15**(6) 2897, (1997).
16. S.Zankovych, T.Hoffmann, J.Seekamp, J.-U.Brunch and C.M.Sotomayor Torres, Nanotechnology **12** 91, (2001).
17. B.Heidari, I.Maximov and L.Montelius, J. Vac. Sci. Technol. B **18**(6), 3557, (2000).
18. L.J.Heyderman, H.Schift, C.David, B.Ketterer, M.Auf der Maur, J.Gobrecht, Microelectronic Engineering **57–58**, 375,(2001).
19. F.Carcenac, C.Vieu, A.Lebib, Y.Chen, L.Manin-Ferlazzo, H.Launois, Microelectronic Engineering **53**, 163, (2000).
20. B.A. Gurovich, D.I. Dolgy, E.A. Kuleshova, E.Z. Meilikhov, A.G. Domantovsky, K.E. Prikhodko, K.I. Maslakov, B.A. Aronzon, V.V. Rylkov, A.Yu. Yakubovsky, Selective Removal of Atoms as a New Method for Fabrication of Nanoscale Patterned Media. Microelectronic Engineering, **69**, N 2-4, pp. 65–75, (2003),
21. Gurovich B.A., Kuleshova E.A., Meilikhov E.Z., Maslakov K.I. Selective removal of atoms as a new method for manufacturing of nanostructures for various applications. Journal of Magnetism and Magnetic Materials 2004, **272–276**, pp. 1629–1630,(2004).
22. Gurovich B.A., Kuleshova E.A., Dolgy D.I. et all. Selective removal of atoms as a new method for fabrication of single-domain patterned magnetic media and multi-layered nanostructures. B. Aktas et al. (eds.) , Nanostructured Magnetic Materials and their Applications. Kluwer Academic Publishers, pp. 13–22,(2004).

Magnetization Reversal Studies of Periodic Magnetic Arrays via Scattering Methods

Arndt Remhof, Andreas Westphalen, Katharina Theis-Bröhl, Johannes Grabis, Alexei Nefedov, Boris Toperverg, Hartmut Zabel

Department of Physics and Astronomy, Institute for Condensed Matter Physics
Ruhr-Universität Bochum, D 44780 Bochum, Germany
arndt.remhof@rub.de

Summary. Magnetic patterns with different shapes and aspect ratios provide control over the remanent domain state, the coercivity, and over different types of reversal mechanisms. We discuss magneto-optical, soft x-ray resonant scattering, and neutron scattering methods for evaluating the vector magnetization during reversal and the higher Fourier components of the magnetization distribution. These scattering methods, providing a statistical averaged signal, contrast real space methods, which give information on individual islands.

1 Introduction

Presently there is a large interest in the fabrication and characterization of magnetic nanostructures [1]. This field is driven by the genuine interest in the magnetic properties of nanostructures as well as by the large potential for applications in magneto-electronic and spintronic devices. The main questions concern (1) the spin structure in nano-patterned media as a function of the shape, size and separation of the islands; (2) the mechanism of reversal via coherent magnetization rotation or domain wall motion; (3) the speed of the reversal and the damping mechanism of spin excitations, (4) different methods for driving the reversal via an external field, spin accumulation, or current torque; (5) the bipolar stability of the spin structure versus thermal fluctuations as a function of size and anisotropy. Advances in instrumental techniques allow to push forward the limits in the spatial and time domain such that magnetic nanostructures can now be fabricated and investigated on the scale of 10–100 nm with a spatial resolution of about 10 nm and on a time scales of pico- to femtosecond.

There are many methods in real space for the investigation of magnetic nanostructures. They will not be discussed here in any detail. The emphasis is on light scattering methods in the visible and in the soft x-ray region, and on neutron scattering. MOKE in the visible range has the advantage of being fast and applicable in the laboratory without the need of a large scale facility for the light source. Furthermore, a number of different MOKE techniques, such as MOKE-microscopy, vector-MOKE, Bragg-MOKE and micro-MOKE, can be utilized, yielding together with micromagnetic simulations a rather detailed picture of the reversal mechanism. With vector-MOKE it is possible to analyze the magnetization vector, which enables to distinguish between different reversal mechanism [2, 19]. In the case of micro-MOKE the hysteresis of single islands is investigated in specular reflection geometry by reducing the cross-section of the incident beam to roughly match their size [20]. In contrast, Bragg-MOKE uses the periodic array of the magnetic features illuminated by a broad incident beam and the hysteresis loops are recorded at different orders of diffraction in off-specular geometry. For ferromagnetic line gratings, the combination of diffraction and the magnetooptical Kerr effect yields information about the mean lateral magnetization distribution [4, 5]. This technique arrives at a natural limit when the period of the array drops below the wavelength of the incident light. Off-specular soft x-ray resonant magnetic scattering (SXRMS) can, however, be employed to overcome this limit [25]. Then smaller feature sizes (or higher orders of diffraction) become accessible. The disadvantage of MOKE in the visible as well as in the x-ray region is the rather small penetration depth of the light wave into the metal, which is about 30 nm. This becomes a problem when studying arrays of magnetic heterostructures or multilayers. Polarized neutron reflectivity (PNR) and polarized neutron scattering (PNS) at small angles is not limited in this respect and a layer resolved vector-magnetometry even for deeply buried layers is feasible [26]. Furthermore, the strength of the PNR method is the off-specular scattering, providing statistical information on correlation lengths and domain fluctuations. However, neutron scattering is intensity limited and this becomes even more severe when studying sample volumes which are severely reduced through the patterning process.

In the following we provide an overview of the different scattering techniques used for the investigation of magnetic patterns. The intention is to describe the current status of experimental techniques without going into theoretical details. The different methods are illustrated by a number of experimental investigations on magnetic nanostructures. We start with a short description of different patterning techniques in Sect. 2. In the experimental Sect. 3 we will first introduce the vector-MOKE technique in Sect. 3.1, followed by a discussion of Bragg-MOKE methods in the visible range (Sect. 3.2) and in the x-ray range (Sect. 3.4). The last part of the experimental Sect. 3.5 discusses polarized neutron methods. In the final Sect. 4 we will summarize and critically compare the different experimental techniques for potential future applications.

2 Patterning

There are generally two routes for the preparation of micro- and nanostructures. The *bottom up* approaches build monodisperse, small particles (2–100 nm) by means of organo-metallic and gas phase synthesis or use of self assembly processes on the atomic scale. *bottom down* techniques on the other hand combines standard thin film deposition methods in combination with lithographic procedures, including novel techniques as diblock copolymer templates, biological templates, alumina templates and nanosphere lithography. For the patterning of extended magnetic arrays usually the *bottom down* route is followed, as self assembly does not produce the required uniformity and lateral order. Fist the desired pattern is defined by a CAD program and written by irradiation into a resist layer. The exposure can be done by electromagnetic radiation (visible, UV or x-ray) or by particles (electrons or ions). While visible or UV lithography generally uses projection methods in which the whole structure is exposed simultaneously through a mask, e-bam lithography is a direct writing technique where the individual elements of the structure are exposed successively. The projection methods allow to prepare rapidly identical structures once the mask is prepared. The resolution of this technique is limited by diffraction. E-beam and ion beam lithography are more versatile and the resolution is limited by the focal spot of the respective beam and the chemical properties of the resist. They are, however, more time consuming. The resist usually consists of a polymer, which undergoes chemical changes upon irradiation. Then the exposed (positive resist) or the unexposed (negative resist) areas of the resist are dissolved in a developer bath, and the structure is transferred into the metallic layer. Using a positive technique, the magnetic film is deposited onto the pre-defined resist template. In a final chemical lift-off step only those islands remain, which have direct contact to the sample or to another metal layer. In the negative technique on the other hand, an etch resistant resist is employed. The magnetic structure is then etched out of a continuous magnetic layer, deposited prior to the resist. In a final step the remaining resist gets removed and only the magnetic islands remain. Both methods are schematically drawn in Fig. 1.

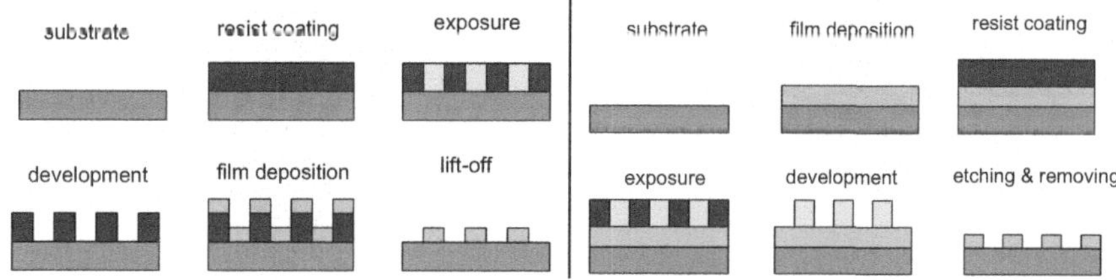

Fig. 1. Patterning steps using positive resist (left panel) and by negative resist (right panel).

3 Experimental Methods for the Analysis of Patterned Magnetic Media

3.1 Vector-MOKE

Magnetic hysteresis measurements are usually performed with a MOKE setup in the longitudinal configuration. This implies the use of s-polarized light, a sample magnetization in the film plane, and a magnetic field applied in the scattering plane and parallel to the film plane. The resulting Kerr angle is then proportional to the component of the magnetization vector parallel to the field direction, $\theta_x \propto m_x$, where m_x is the longitudinal component of $\boldsymbol{M}$ projected parallel to $\boldsymbol{H}$. This kind of measurement can not distinguish between a magnetization reversal via domain rotation and/or via domain formation and wall motion. It is therefore advantageous to measure not only the component of the magnetization along the applied field, but also in the orthogonal direction in order to reconstruct the magnetization vector $\boldsymbol{M}$ from the measurement. The longitudinal MOKE can be used as a vector-magnetometer, if, in addition, the external magnetic field is applied perpendicular to the scattering plane and the sample is simultaneously rotated by 90° with respect to the scattering plane, keeping the rest of the setup constant. In this perpendicular configuration, MOKE detects the magnetization component parallel to the scattering plane and perpendicular to the magnetic field, $\theta_y \propto m_y$, as has been shown by [2].

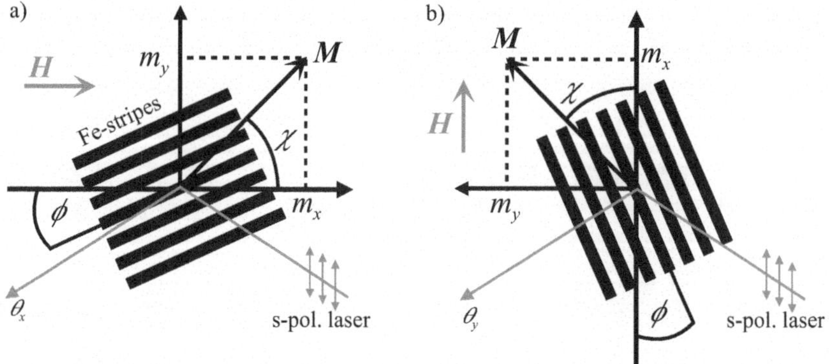

Fig. 2. Definition of the sample rotation ϕ and the angle χ of the magnetization vector $\boldsymbol{M}$ for the case of the longitudinal setup (a) and the perpendicular setup (b). In order to measure the y-component of the magnetization m_y the field and the sample are rotated by 90°, such that the angle ϕ is held constant, but the magnetization component m_y is in the scattering plane.

Vector-MOKE has originally been applied to continuous magnetic films [2]. Here it is of interest to utilize this method for the analysis of magnetic patterns. The geometry of the setup for the case of a stripe pattern is sketched in

Fig. 2. ϕ defines the the angle between the stripes (easy axis) and the applied field direction. χ is the angle between the magnetization vector $\boldsymbol{M}$ and the applied field. After measuring a hysteresis in the configuration sketched in Fig. 2(a), the sample and the field is rotated by 90° (Fig. 2(b)), thus leaving the relation between stripes and field constant. The original and the rotated configuration combined yield the average vector magnetization for one particular angle ϕ. Obviously ϕ has to be changed in consecutive runs to probe the reversal mechanism for any angle between the easy and the hard axis.

The formalism of vector-MOKE is independent of the patterning as the specularly reflected beam is used for the MOKE signal. The specular signal averages over the magnetic islands as well as over the materials in the interspace. This will be different in the next section, where we investigate the off-specular Bragg-diffracted MOKE signal. Both components, m_x and m_y, yield the vector sum for the average magnetization vector $\boldsymbol{M}$ sampled over the region, which is illuminated by the laser spot. This area is approx. 1 mm^2. The magnetization vector can be written as

$$\boldsymbol{M} = \begin{pmatrix} m_x \\ m_y \end{pmatrix} = |\boldsymbol{M}| \begin{pmatrix} \cos\chi \\ \sin\chi \end{pmatrix} . \tag{1}$$

The proportionality constant between the Kerr angles θ_x and θ_y and the two magnetization components is *a priori* unknown. However, there are good reasons to assume that they are equal. In this case one can write:

$$\frac{m_x}{m_y} = \frac{\cos\chi}{\sin\chi} = \frac{\theta_x}{\theta_y}, \tag{2}$$

from which follows the rotation angle of the magnetization vector:

$$\chi = \arctan\left(\frac{\theta_y}{\theta_x}\right) . \tag{3}$$

Furthermore one can express $|\boldsymbol{M}|$, normalized to the saturation magnetization:

$$\frac{|\boldsymbol{M}|}{|\boldsymbol{M}^{sat}|} = \sqrt{\left(\frac{\theta_x}{\theta_{sat}}\right)^2 + \left(\frac{\theta_y}{\theta_{sat}}\right)^2} . \tag{4}$$

The results of vector-MOKE measurements provide important information on the type of magnetization reversal. Two limiting cases can easily be distinguished (see Fig. 3):

- Coherent rotation (Fig. 3(a)) occurs when the length of the magnetization vector $|\boldsymbol{M}|$ is constant during the reversal, while the magnetization vector $\boldsymbol{M}$ rotates from one direction into the other. Accordingly, the y-component m_y, plotted as a function of the x-component m_x, increases with decreasing longitudinal component, and exhibits a maximum at remanence.

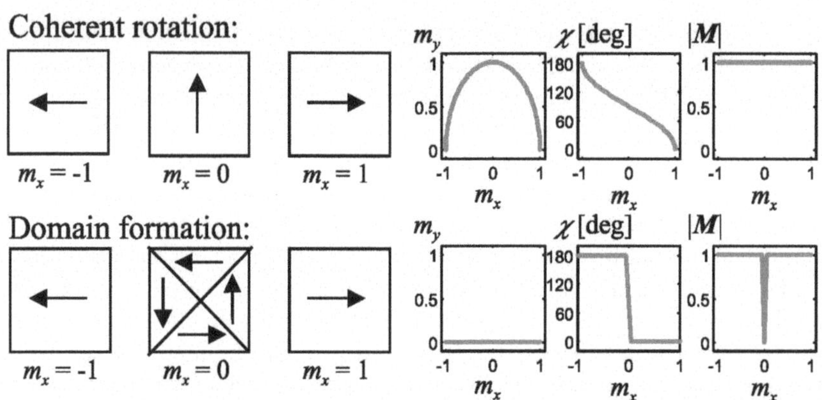

Fig. 3. Two limiting cases of magnetization reversal and the resulting vector-MOKE measurements can be distinguished. The top panel shows the case of coherent rotation. The reversal is sketched and the y-component, the angle χ of rotation, and the magnitude of the magnetization $\boldsymbol{M}$ are plotted as a function of x-component of the magnetization vector. Bottom panel depicts the case of domain formation and without rotation. The same quantities are plotted as in top panel.

– Domain formation (Fig. 3(b)) is recognized by a magnetization vector, which remains aligned with the external field, but changes its magnitude from negative to positive saturation and vice versa. In this case the y-component stays zero for all field values. If no y-component can be detected, no rotation of the magnetization takes place and the reversal is governed by domain processes.

Examples for vector-MOKE

In the following we discuss the magnetization reversal of a stripe pattern using vector-MOKE [3]. The stripes consist of a 20×10 mm^2 and 90 nm thick polycrystalline $Co_{0.7}Fe_{0.3}$ film grown by DC magnetron sputtering and patterned subsequently by optical lithography. The magnetic anisotropy is dominated by the shape anisotropy with no further intentional anisotropy. The stripes have a width of $w = 1.2$ μm and a grating parameter of $d =$ 3 μm as can be seen from the atomic force microscope (AFM) topograph reproduced in Fig. 4. Due to the high aspect ratio, the easy axis in this pattern is aligned parallel to the stripes and in remanence the stripes are in a single domain state.

The left row of Fig. 5 shows four typical longitudinal MOKE hysteresis loops taken from the CoFe stripes with different in-plane angles ϕ displaying the x-component of the magnetization vector. The hysteresis loop in (a) corresponds to an external magnetic field oriented parallel to the stripes. In this case, we find an almost square hysteresis loop, which represents the typical behavior of a sample when magnetically saturated parallel to the easy axis of

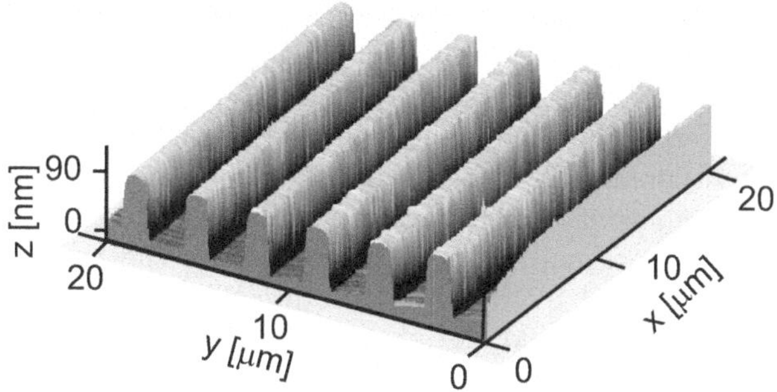

Fig. 4. Surface topography of the array of $Co_{0.7}Fe_{0.3}$ stripes obtained with an atomic force microscope shown in a 3-dimensional surface view. The displayed area is 20×20 μm^2.

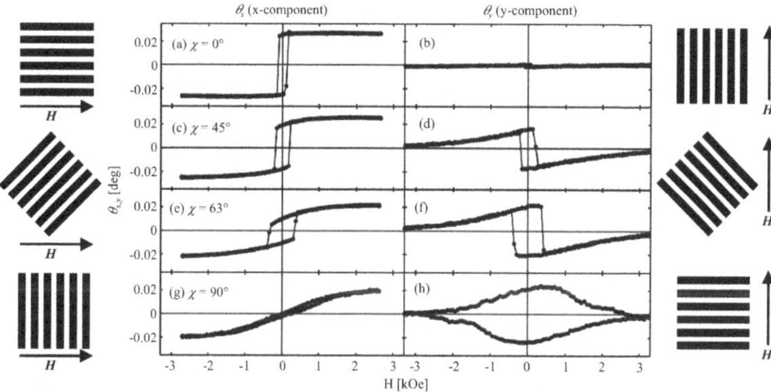

Fig. 5. MOKE hysteresis loops measured the x-component (left figures) and the y-component (right figures) of the magnetization vector. For both configurations in (a) and (b) the external magnetic field is oriented parallel to the stripes (easy axis configuration); in (c) and (d) the stripes are rotated by 45° and in (e) and (f) by 63° with respect to the direction of the external field; in (g) and (h) the magnetic field is oriented perpendicular to the stripes (hard axis configuration).

the magnetization. The coercive field is $H_c = 140$ Oe. The coercive field increases to $H_c = 200$ Oe and $H_c = 320$ Oe for intermediate angles of rotation of $\phi = 45°$ (c) and 63° (e), respectively. Figure 5(g) shows the corresponding hysteresis loop for the CoFe stripe array oriented perpendicular to both the external field and the plane of incidence. Here, a typical hard axis hysteresis loop is obtained. The saturation field measured with MOKE in the hard axis configuration exceeds 1000 Oe.

The y-component of the magnetization vector is reproduced in the right column of Fig. 5. The hysteresis loop reproduced in (b) was determined for a magnetic field parallel to the CoFe stripes, i.e. parallel to the easy axis.

Ideally, within this configuration the measured Kerr rotation should remain zero unless components of the magnetization lie in the plane of incidence during the magnetization reversal process. As can be seen from Fig. 5(b), the measured Kerr rotation is indeed almost zero, thus no rotation of the magnetization occurs in this configuration. Figure 5(d) and (f) show the corresponding y-component of the magnetization vector measured for higher angles of rotation ϕ. Finally Fig. 5 (h) reproduces the hysteresis loop for a magnetic field parallel to the hard axis direction, i.e. in a direction perpendicular to the stripes and the plane of incidence. In this configuration the y-component of the magnetization vector always reflects a rotation of the magnetization away from the magnetic field.

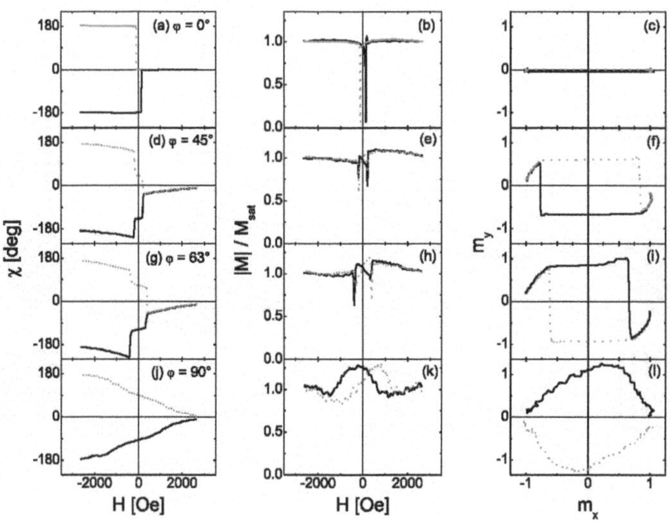

Fig. 6. Plot of the angle of rotation χ (left panel), magnitude of the magnetization vector $|\boldsymbol{M}|$ (middle panel), and y versus x-component of the MOKE signal for different sample rotation angles ϕ according to the measurements shown in Fig. 5. Solid lines are for ascending magnetic fields and dotted lines for descending fields.

From the x and y-components the magnetization vector can be reconstructed as discussed further above. This is shown in Fig. 6, where the angle of the magnetization vector χ and the magnitude $|-\boldsymbol{m}|$ is plotted as a function of the external field for all four stripe orientations. The analysis confirms that for a magnetic field direction parallel to the stripes the reversal proceeds by nucleation and domain wall motion within a very narrow region around the coercive field, H_c (see Fig. 6(a,b)). However, for field directions $0 < \phi < 90°$ there is a sizeable contribution from coherent rotation up to the coercive field, where some domains are formed (see Fig. 6 (d-i)). For $\phi = 90°$ the reversal mechanism appears to be entirely dominated by coherent rotation of the magnetization vector and no switching takes place. In the hard axis direction the magnetization vector describes a complete 360° rotation during the full magnetization cycle without any discontinuity (see Fig. 6 (j,k,l)). For other directions a switching of the magnetization by 180° is observed at H_c, which can be viewed as a head-to-head domain wall movement through the stripes. These conclusions have been confirmed by imaging domains within the stripe pattern via Kerr microscopy [3].

3.2 Bragg-MOKE

Whenever a lateral magnetic structure has a periodicty of the order of the wavelength of the illuminating laser light, the sample acts as an optical grating, leading to interference effects in the reflected laser light. Kerr hysteresis loops can be measured both in specular reflection and at the diffraction spots. The measurement and analysis of magnetic hysteresis curves from interference spots of higher order has been dubbed *diffraction-MOKE* [4] or *Bragg-MOKE* [5].

For Bragg-MOKE measurements the angle of incidence α_i should be set to 0° (perpendicular incidence) in order to produce a symmetric diffraction pattern for positive and negative orders of interference. Furthermore, in case of a stripe arrays the grating has to be set with the stripes perpendicular to the scattering plane and the external magnetic field, i.e. the Bragg-MOKE hysteresis curves are obtained in the hard axis configuration only. Any other stripe orientation moves the scattering plane away from horizontal, requiring a special detector arrangement. The geometry of the Bragg-MOKE measurements is sketched in Fig. 7.

Examples for Bragg-MOKE measurements

An illustrative example is provided by a thin polycrystalline Fe film on a sapphire substrate, which has been laterally structured into stripes of different widths (w = 0.5, 2.1, 2.5, and 3.7 μm) but constant grating parameter (d = 5 μm). The thickness of 50 nm is also the same for all stripe arrays. The magnetic anisotropy is dominated by the shape anisotropy, while the polycrystalline film averages the intrinsic magneto-crystalline anisotropy.

The Bragg-MOKE results for the different gratings are reproduced in Fig. 8 [5]. Because of the symmetric scattering geometry, only results from positive orders of diffraction are shown. The rows of Fig. 8 display the Bragg-MOKE hysteresis curves for one stripe width at different orders of diffraction $n = 1 - 3$. The columns represent the same order of diffraction for different Fe stripe widths w. Obviously, the shape of the hysteresis curves changes with the order of diffraction. While the first order hysteresis curves appear normal, anomalies are evident at higher orders.

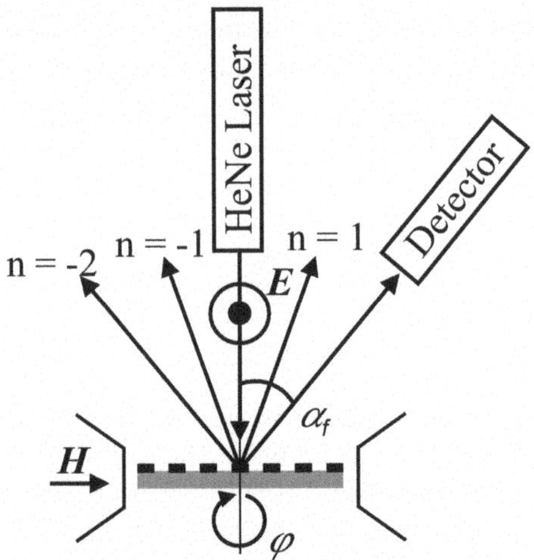

Fig. 7. Geometry of the Bragg-MOKE setup. The s-polarized light in perpendicular incidence ($\alpha_i = 0$) is diffracted from the grating. The sample rotation angle ϕ is selected, such that the magnetic field is perpendicular to the stripes. The Kerr detector can be rotated perpendicular to the plane of incidence by an angle α_f.

In Fig. 9 MOKE measurements are shown for the same stripe arrays in specular geometry, $n = 0$, and for stripe orientations parallel (easy) and perpendicular (hard) to the applied field in the scattering plane. A comparison between specular and Bragg-MOKE is only possible for the hard axis orientation, since the diffraction spots need to be in the scattering plane. Aside from the stripe pattern with the smallest width, the zeroth and the first order hysteresis curves look very much alike. In the specular configuration the substrate contributes to the MOKE signal in an uncontrolled fashion. This is particularly problematic for arrays with a low coverage of the substrate. On the other hand, off-specular Bragg-MOKE hysteresis curves may not be representative for the "true" hysteresis, due to inference effects and angle dependent magneto-optical parameters, to be discussed further below. However, according to our experience and for arrays of simple structures the $n = 1$ and $n = 0$ hysteresis loops usually have a similar shape with the added advantage that the former has a much better signal to noise ratio.

In the specular MOKE measurements no anomalous hysteresis loops are observed. There are no contributions from higher order MOKE effects. But in the Bragg-MOKE curves in Fig. 9 one finds several hysteresis loops exhibiting an anomalous shape. The anomalies displays a symmetry with respect to the origin of the diffraction order $n = 0$, which is not consistent with second-order effects [6]. In Bragg-MOKE studies for ferromagnetic dot arrays [7] and anti-dot arrays [6, 8] similar anomalous hysteresis loops have been observed. The effect has been shown to be caused by the magnetic domain structure within the patterned films.

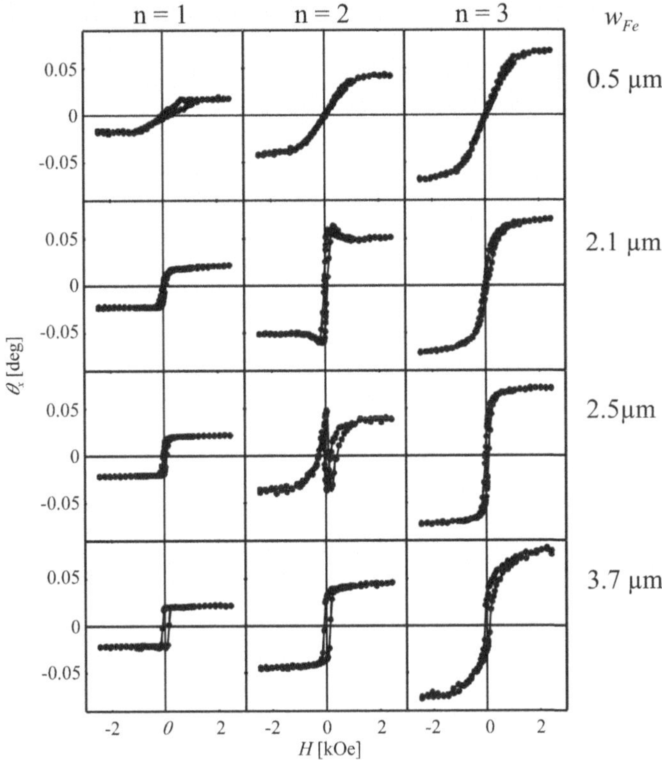

Fig. 8. Bragg-MOKE hysteresis curves from the polycrystalline Fe gratings. Each line in the figure represents the measurements of one grating with constant w, as indicated in the figure. The stripe width w increases from top to the bottom. In the columns the positive first three orders of diffraction are shown.

In Fig. 10(a) a Kerr microscopy image is shown of the domain pattern taken from the stripe array with $w = 2.5$ μm in the demagnetized state. We observe a very regular domain pattern with closure domains at the stripe edges, as depicted schematically in Fig. 10(b). In the remanent state (not shown) one essentially observes similar domains with one magnetization direction in the interior of the stripes. The essential idea of Bragg-MOKE is – in analogy to regular x-ray or neutron Bragg-diffraction – that the different orders of diffraction represent the Fourier components of the magnetization distribution within the sample and during the magnetization reversal, from which some features of the domain pattern can be reconstructed.

In the past the Bragg-MOKE hysteresis loops have been analyzed using so-called *diffraction hysteresis loops* (DHL). In this representation the Bragg-MOKE signal is plotted as a function of the normalized magnetization, the latter one being determined from the specular measurements. The DHL are then modelled by assuming a particular domain structure. As pointed out

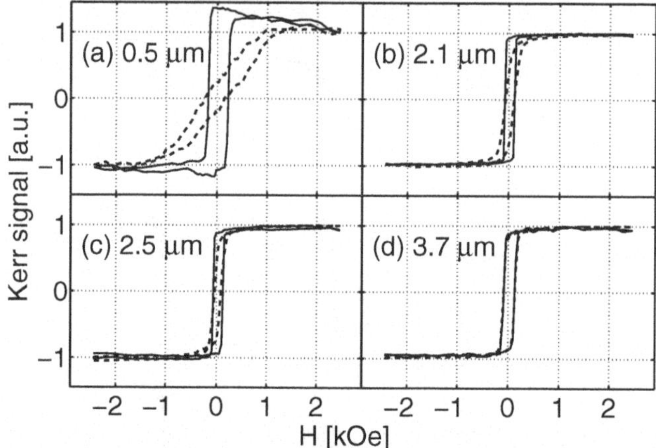

Fig. 9. MOKE hysteresis loops measured along and perpendicular to the stripes for the gratings with (a) w = 0.5 μm, (b) w = 2.1 μm, (c) w = 2.5 μm, and (d) w = 3.3 μm. The hard axis (dashed line) is measured with the external field perpendicular to the stripes and the easy axis (straight line) with the field parallel to the grating.

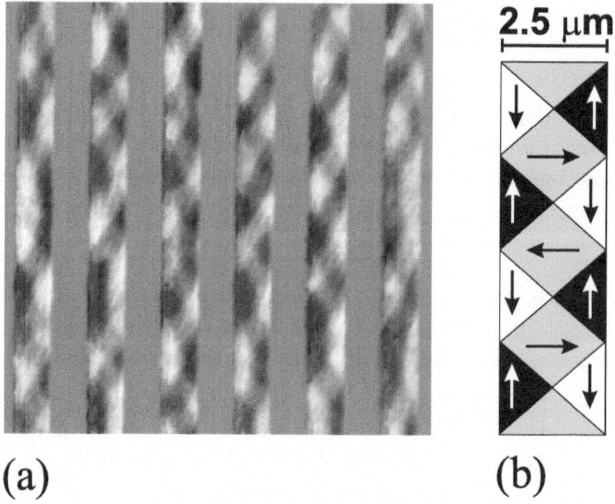

Fig. 10. (a) Kerr microscopy image in the demagnetized state of a Fe grating with the lattice parameter d = 5 μm and the stripe width w = 2.5 μm. The field direction during demagnetization was perpendicular to the stripes. (b) Orientation of the magnetization within the domains schematically.

especially in Ref. [6, 7], the magnetic signal obtained from hysteresis measurements at diffraction spot n represents the nth Fourier component of the magnetization distribution. With simple one-dimensional models the DHL for several typical domain structures can be calculated. All models assume the magneto-optical signal to be given by the real part of the Fourier transform:

$$f_n^m = \int_0^d m(x) \exp(ikx) dx. \tag{5}$$

Here d is the grating parameter, $m(x)$ is the magnetization distribution, and the wave vector is given by $k = 2\pi n/d$.

Instead of plotting and computing DHL for particular domain structures, it has now become possible to take the Fourier transform directly from simulated magnetization distributions [9, 10], using one of the standard micromagnetic modelling packages, such as OOMMF [11]. We discuss this method for the case of an Fe stripe array with alternately thick (2 μm) and thin (1 μm) bars, both having the same length of 30 μm. In one direction, the grating period is 6 μm, in the other direction 40 μm. Details of the structure are shown in Fig. 11 [12].

The Kerr hysteresis loops were obtained in hard axis orientation. The Kerr hysteresis loop measured at the specular intensity spot shows a typical hard axis behavior with vanishing remanence. The hysteresis measured at the second diffraction order is similar to the specular one. For the first and the third order of diffraction the shape of the hysteresis loops is completely different.

To explain the behavior of the hysteresis loops measured at the diffraction spots, micromagnetic calculations have been used [12]. For the simulations we consider two rectangles of the structure having dimensions 2 μm × 30 μm and 1 μm × 30 μm, respectively, and a separation between the rectangles of 1.5 μm with a respective thickness of 100 nm. The crystalline anisotropy constant K_1 was set to 0. To compare the simulations with the Kerr hysteresis loops measured at the diffraction spots, the magnetic form factor of order n, f_n^m, has to be calculated for each field value and as a function of the magnetization distribution inside the stripe. Taking the real and imaginary part of the magnetic form factor, measured data have been fit with the expression:

$$m_n = A_n \Re(f_n^m) + B_n \Im(f_n^m), \tag{6}$$

where A_n and B_n are adjustable parameters. In Fig. 12 the measured and calculated data are compared. For all orders of diffraction the calculated data match the measured hysteresis loops rather well. In Fig. 13 four magnetization profiles for increasing magnetic field values are presented. The remagnetization process starts at the longer edges of the rectangles. The smaller rectangle (1 × 30 μm^2) breaks up into domains for field values around –1000 Oe. The bigger rectangles follow at –600 Oe. For the bigger rectangle the remagnetization process is almost finished at 900 Oe, while the smaller one is still in a domain state.

Fig. 11. Topography image of the stripes as obtained by atomic force microscopy.

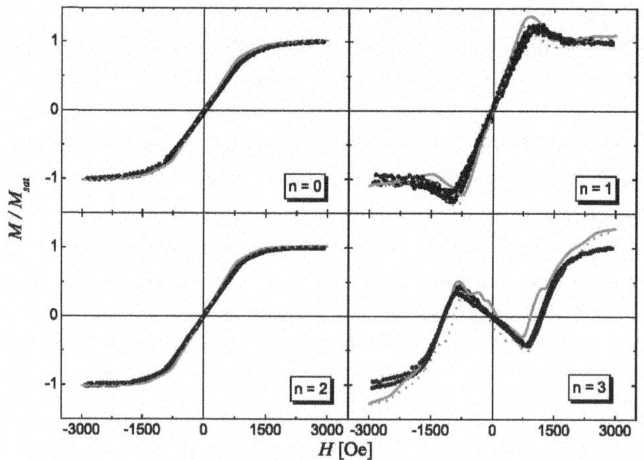

Fig. 12. Kerr hysteresis loops from the array measured at different orders of diffraction (dots) and calculated data from micromagnetic simulations (lines).

The combination of Bragg-MOKE results with micromagnetic calculations is a very good tool for investigating the magnetization in periodic arrays of magnetic microstructures. The good agreement between Bragg-MOKE data and calculated hysteresis loops of n-th order provides confidence that the hysteresis loops measured in higher order contain meaningful informa-

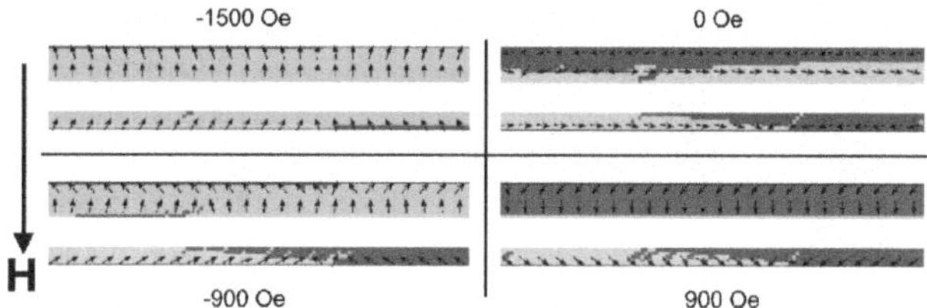

Fig. 13. Calculated magnetization profiles from micromagnetic simulations.

tion. In fact, in case of periodic lateral magnetic structures it is not sufficient to determine only the 0th order specular MOKE hysteresis.

3.3 Combination of Vector and Bragg-MOKE

Vector and Bragg-MOKE techniques are no separate techniques which work only alone. Both can be combined. Since Vector-MOKE requires the rotation of sample and applied field by 90°, only patterns on a square lattice can be analyzed with this combined technique. First results have now been obtained from a pattern of equilateral triangles [18]. A polycrystalline Fe film (thickness = 30 nm) was deposited at room temperature on pre-defined resist templates by ion beam sputtering. The resist templates were fabricated by conventional electron beam lithography. The grating parameter in x and y-direction was 12 μm, the length of the sidepieces 6 μm. In Fig. 14 the Bragg-MOKE hysteresis measured at the first and second diffraction spot for the x- and y-component are compared with the results from micromagnetic simulations. The agreement is obvious. From the magnetization profiles obtained from OOMMF the devolution of the remagnetization process can be concluded. When the field reaches the point A (cf. Fig. 14), first domains with magnetization parallel to positive field direction are formed in the sidepiece which is parallel to the field. The forming of domains in the other sidepieces is visible in point B in the hysteresis. In point C the magnetization reversal is almost finished. With increasing field the magnetization rotates into saturation.

The combination of both techniques is certainly a very powerful method for detailed analysis of the magnetization reversal of patterned magnetic materials on the micro- to submicrometer scale.

Further investigations with Bragg and Vector-MOKE

The Bragg-MOKE effect has been used to study domain formations during the magnetization reversal of a variety of different sample geometries and shapes. Circular holes, elliptic holes, and square holes have been investigated [6, 8, 14] as well as circular islands and square rings [15]. Of particular interest is the investigation of vortex states in circular and elliptic islands

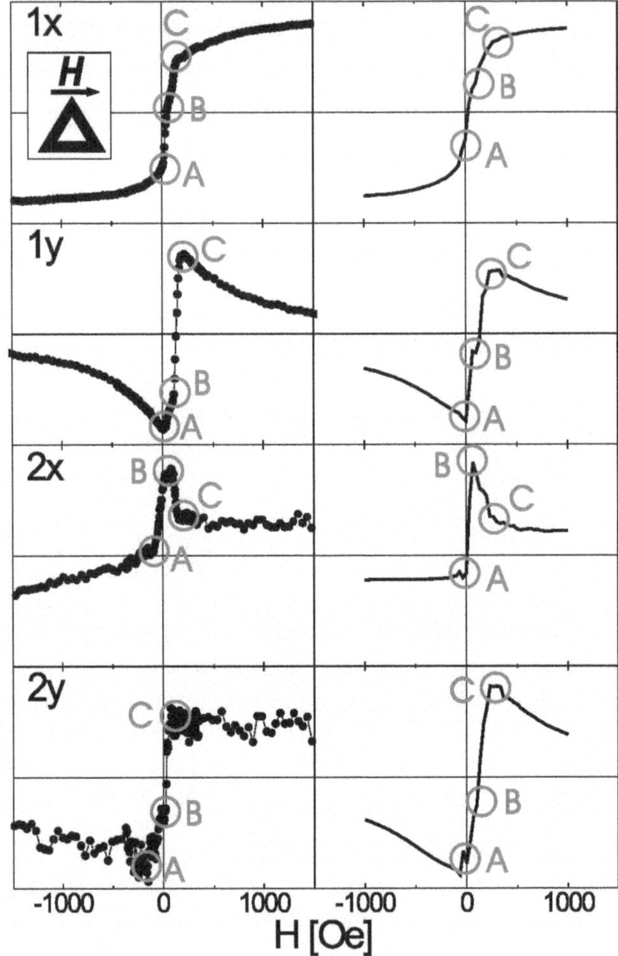

Fig. 14. Measured (left column) and calculated (right column) Bragg-MOKE hysteresis loops for the x- and y-component at the two first diffraction spots; only the part of the loops coming from negative to positive field values are shown. The inset shows the structure geometry.

and how these states effect the shape of the hysteresis in higher order of diffraction [9, 10]. More recently the exchange interaction between Co and Fe stripes separated by a Cr interlayer [13], the dipolar interaction between magnetic islands of different shape have been investigated via Bragg-MOKE [16], and the exchange bias effect of a CoFe/MnIr bilayer could be explained with a Spin Glass model comparing the simulations with Vector-MOKE data [17].

3.4 Soft x-ray Resonant Magnetic Scattering

The magnetic sensitivity of soft x-rays originates from the difference between the absorption of left and right circularly polarized x-rays for a magnetic material [21, 22]. This x-ray magnetic circular dichroism (XMCD) is widely used to investigate element selective hysteresis curves, the dynamics of magnetization reversal on the nanosecond scale and the imaging of magnetic domains with a photo emission electron microscope (PEEM). Soft x-ray resonant magnetic scattering (XRMS) yield the combined element specific spectroscopic information originating from XMCD together with the depth resolved structural information as measured in conventional (non resonant) small angle x-ray diffraction, such as density profiles and interface roughness in magnetic heterostructures [24, 23]. A periodically, lateral structured sample will act as an optical grating and generate off-specular diffraction spots. Thus, similar to Bragg-MOKE in the visible range, the magnetization reversal of patterned magnetic arrays can be studied with off-specular x-ray resonant magnetic scattering [27, 28, 29]. In this case the magnetic dichroism of circularly polarized x-rays with energies close to the L-absorption edges of 3d transition metals is exploited to determine the magnetic hysteresis in the diffraction mode.

In general, the differential cross section for x-ray scattering in the kinematic approximation is given by

$$\frac{d\sigma}{d\Omega} = \left| \int_V d^3 r N f(\boldsymbol{q},\omega) \exp\left(-i\boldsymbol{q}\cdot\boldsymbol{r}\right) \right|^2 , \tag{7}$$

where $f(\boldsymbol{q},\omega)$ is the atomic scattering amplitude and N is the number density of scatterers.

If in the resonant part only dipole transitions are considered, the complete scattering amplitude for non-resonant and resonant scattering is given by [30]

$$f(\boldsymbol{q},\omega) = (\boldsymbol{\epsilon}_f^{*}\cdot\boldsymbol{\epsilon}_i)(-r_e Z+F^{(0)})+i(\boldsymbol{\epsilon}_f^{*}\times\boldsymbol{\epsilon}_i)\cdot\boldsymbol{m}F^{(1)}+(\boldsymbol{\epsilon}_f^{*}\cdot\boldsymbol{m})(\boldsymbol{\epsilon}_i\cdot\boldsymbol{m})F^{(2)}, \tag{8}$$

with

$$F^{(0)} = \frac{3\lambda}{8\pi}\left[F^1_{-1} + F^1_1\right], \tag{9}$$

$$F^{(1)} = \frac{3\lambda}{8\pi}\left[F^1_{-1} - F^1_1\right], \tag{10}$$

$$F^{(2)} = \frac{3\lambda}{8\pi}\left[2F^1_0 - F^1_{-1} - F^1_1\right]. \tag{11}$$

Here the $F^L_M(\omega)$ are the transition matrix elements for dipole transitions, and $\boldsymbol{\epsilon}_i$ and $\boldsymbol{\epsilon}_f$ are polarization vectors of the incident and scattered radiation, respectively. The unit vector $\boldsymbol{m}$ points along the magnetization direction, which defines the quantization axis of the system. Terms proportional to $\boldsymbol{\epsilon}_f^{*}\cdot\boldsymbol{\epsilon}_i$ describe non-resonant and resonant charge scattering. The term involving $(\epsilon_f^* \times \epsilon_i) \cdot \boldsymbol{m}$ is first order in the magnetization and yields circular dichroism and Kerr effects. The term proportional to $(\epsilon_f^* \cdot \boldsymbol{m})(\epsilon_i \cdot \boldsymbol{m})$ is second order in the magnetization and causes linear dichroism or the Voigt effect, which is only relevant for the transmission geometry and will not be considered further on.

Element specificity of the scattering cross section is obtained by tuning the x-ray energy to the appropriate absorption edge. In 3d transition metals the excitation of $2p_{1/2}$ and $2p_{3/2}$ electrons into unoccupied 3d states leads to strong absorption edges with energies in the soft x-ray region, referred to as L_{II}- and L_{III}-edges, respectively [31]. Resonant scattering causes a large enhancement of the scattering cross-section. The intensity difference between left and right circularly polarized x-rays is a fingerprint for the element and simultaneously for the existence of a local magnetic moment of that element. The scattering geometry for the application of resonant x-ray magnetic scattering is sketched in Fig. 15 [32]. The magnetic field is applied parallel to the scattering plane in a geometry similar to the longitudinal MOKE effect described in Sect. 3.1. Instead of changing the chirality of the photons, in practice the sample is exposed to a magnetic field. Positive and negative saturation is equivalent to a change of the polarization. Furthermore, the possibility to tune the photon energy around the absorption edges enables independent measurements of the structural order (at arbitrary energies) and the magnetic order (at the resonant energy). A special chamber (ALICE) was built to fulfill the geometric conditions for xrms measurements, including a large parameter space in magnetic field and temperature [33].

The structural diffraction pattern exhibits satellite peaks around the central specular reflection. The application of an in-situ magnetic field at photon energies near the absorption edges allows us to monitor the magnetic contribution of each individual satellite peak through the whole hysteresis loop from one saturated state to the other one. While the hysteresis loop, recorded at the central specular reflection, yields the average magnetization of the sample, the ones recorded at higher orders yield the Fourier components of the magnetic form factor of the ferromagnetic islands during the reversal process.

Examples for resonant x-ray scattering from stripe arrays

In the following we will discuss results obtained at the off-specular diffraction peaks of a magnetic array, consisting of an array of rectangular permalloy (Py) islands with lateral size of 0.3 μm × 3 μm and thickness of 25 nm, set in a square grid with a periodicity of 5 μm. A secondary electron micrograph of the Py array is shown in Fig. 16. The aspect ratio and the thickness of the individual Py islands were chosen to create single domain particles [34], which was confirmed by magnetic force microscopy images at remanence. For the resonant x-ray measurements a photon energy of 852 eV (close to the Ni edge) and an incident angle close to 4° was chosen. Figure 17 depicts the two respective off-spectular q_x scans, recorded in-situ in an applied magnetic

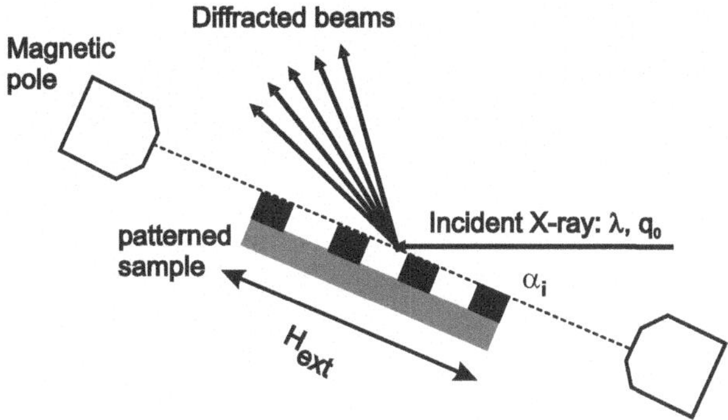

Fig. 15. Scattering geometry for the soft-x-ray experiments using the ALICE difractometer at the synchrotron im Berlin.

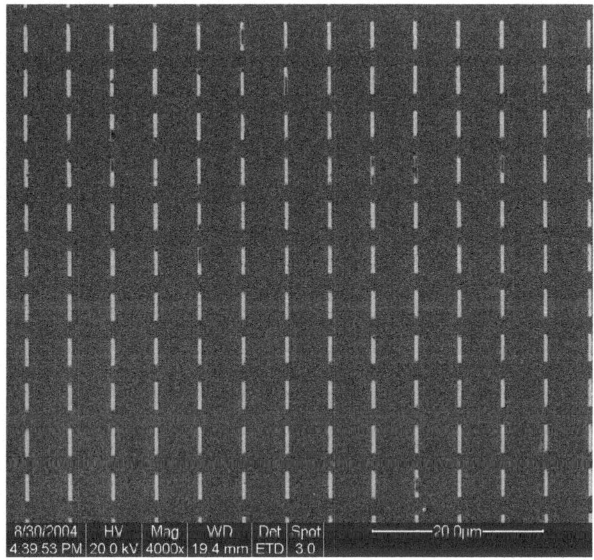

Fig. 16. SEM micrograph of the permalloy array. Rectangular islands of lateral size of 0.3 μm × 3 μm are set in a square grid with a periodicity of 5 μm.

field of ±1500 Oe along the short axis of the rectangular islands, respectively. The specular reflection at $q_x = 0$ is accompanied by two Yoneda wings at $q_x = \pm 0.03$ nm^{-1} and by numerous equally spaced diffraction peaks originating from the regular Py pattern. The magnetic contrast at those satellite reflection can clearly be seen. Off-specular hysteresis loops could be measured up to the 35th order of diffraction. A selection of those are displayed in Fig. 18. All hysteresis loops exhibit a S-like shape with the same coercive field of 35 Oe, the same remanence of 0.3M_{sat} and a saturation of 400 Oe.

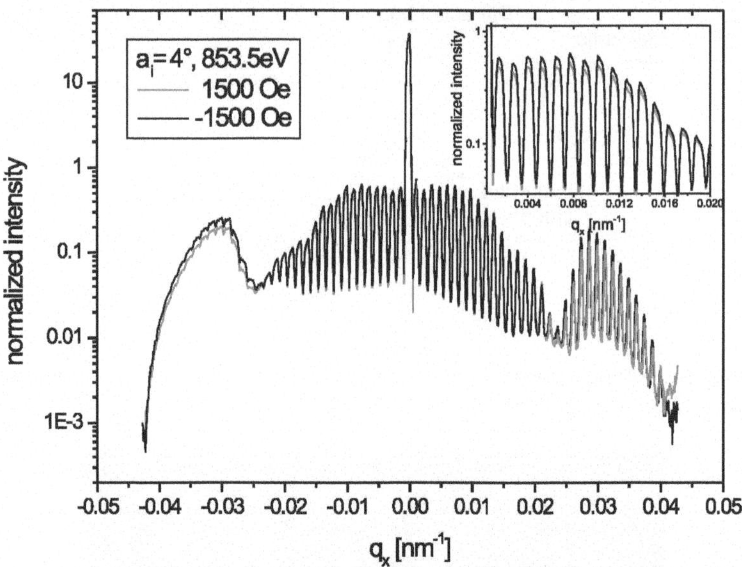

Fig. 17. Q_x scans recorded in both magnetic saturations. An incident angle of 4° and at a photon energy of E = 852 eV (Ni-edge) was chosen. The specular reflection at $q_x = 0$ is accompanied by more than 30 equally spaced satellite reflections. The magnetic asymmetry reaches up to 25%. The inset shows a detail.

Unlike the hysteresis loops recorded with visible light, there is no change in sign of the hyteresis loops is observed for negative orders of diffraction.

As discussed before for the optical Bragg-MOKE, the magnetic signal obtained from hysteresis measurements at diffraction spot n represents the nth Fourier component of the mean magnetization distribution. In the kinematical approach it yields the magnetic form factor of a single island, analogous to eqn (5).

As the Py islands were prepared to be single domain particles we expect a coherent rotation of the magnetic moments towards the easy axis given by the shape anisotropy of the islands, i.e. parallel to the long axis of the islands. This results in uniform decrease of the projection of the magnetization to the scattering plane, leading to order-independent hysteresis loops. This has indeed been observed.

Consequently, in this case the magnetic hysteresis loops recorded at higher order off-specular Bragg reflections are representative for the whole array. Therefore, the magnetic signal measured at e.g. the firt order Bragg reflection filters out the magnetic behavior the stripe array under investigation and suppresses possible contributions from unpatterned parts of the sample as they do not contribute to the interference process.

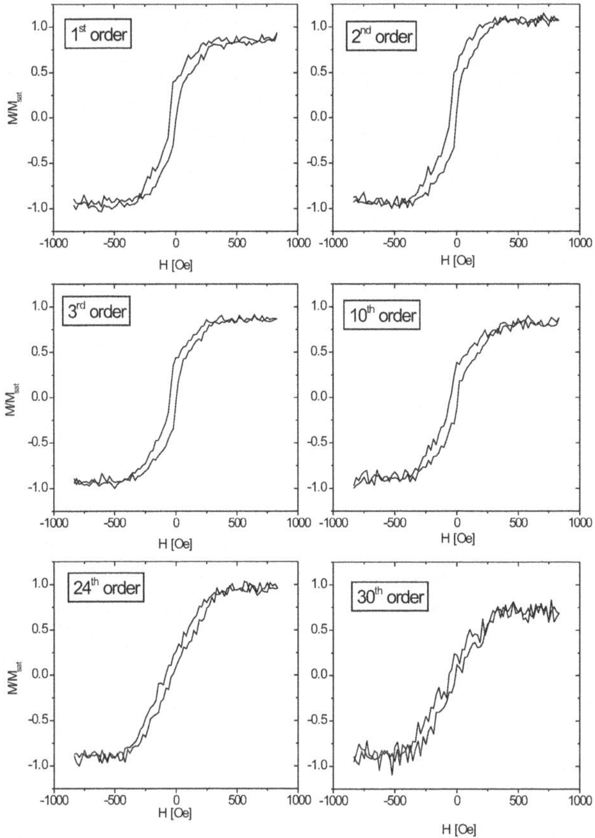

Fig. 18. Selected Off-specular hysteresis loops recorded at the corresponding diffraction peaks of Fig. 17. An incident angle of 4° and a photon energy of 852 eV (close to the Ni edge) was chosen.

Further results and comparison with other techniques

Patterned magnetic samples can not only be produced by artificial lithographic means, but also by self-organized minimization of the magnetic field energy. For instance, in FePd thin-film samples the competition between perpendicular magnetic anisotropy and shape anisotropy leads to the formation of highly ordered stripe domain patterns with a magnetization component perpendicular to the film plane. The magnetic stripes with a period of 100 nm give rise to purely magnetic peaks in the diffraction pattern [35]. Although magnetic force microscopy provides a real space image of the stripe pattern, only with scattering experiments the coherence length can be determined.

Specular and off-specular x-ray scattering has also been used to follow the field dependent magnetization in an array of 1×0.35 μm permalloy rectangles, which is similar to the shape describe above [29]. Hysteresis measurements

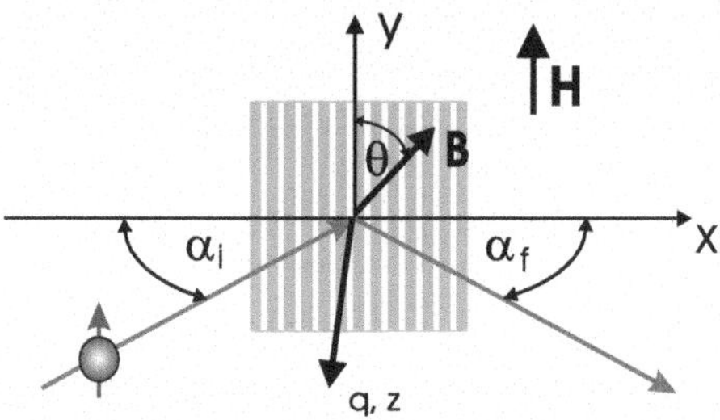

Fig. 19. Scattering geometry for polarized neutron reflectivity studies of a magnetic stripe array. The external field $\boldsymbol{H}$ is applied parallel to the y-axis, along which the neutrons are polarized. For specular reflectivity studies the scattering vector $\boldsymbol{q}$ is parallel to the z-axis. χ is the angle of the sample rotation with respect to the applied field (same definition as in Fig. 2). The magnetic induction $\boldsymbol{B}$ makes the angle θ with respect to $\boldsymbol{H}$. α_i and α_f refer to the incident and exit angles of the neutrons to the sample surface

were performed parallel and perpendicular to the easy axis and a strong magnetic anisotropy induced by the patterning, with a magnetic hard axis in the direction of the shorter side of the rectangles was observed, in agreement with measurements of [32].

Even though the theory of the off-specular resonant x-ray scattering from laterally structured ferromagnetic samples is not fully understood, it is reasonable to assume that one can interpret the hysteresis loops like the corresponding Bragg-MOKE loops. Soft x-rays have several advantages over visible light. Due to their shorter wavelength higher orders of diffraction become accessible: for the Py sample discussed above the magnetic signal could be obtained for more than 30 orders. X-rays also allow to study structures that are much smaller than the wavelength of visible light. The greatest advantage is the element selectivity, which allows to address a specific material within a magnetic heterostructure. The disadvantage of both techniques in the visible as well as in the x-ray region is the rather small penetration depth of the light wave into the metal which hinders the study of thick films, buried layers or multilayers. This limitation could be overcome by using off-specular polarized neutron reflectivity, which will be discussed in the following chapter.

3.5 Polarized Neutron Reflectivity and Scattering

Neutron scattering from nanostructured magnetic arrays is a challenging task. Nevertheless it is worth pursuing this task because of the unique information neutron scattering can offer with respect to correlation effects and domain fluctuations.

A few basic properties of polarized neutron reflectivity shall be recalled here for later use. For more details we refer to [26, 37]. The experimental set-up is somewhat similar to the MOKE set-up, as schematically shown in Fig. 19. The magnetization vector $\boldsymbol{M}$ is assumed to be in the sample plane (in-plane anisotropy) and is perpendicular to the scattering vector $\boldsymbol{Q}$ at specular reflection. Furthermore, $\boldsymbol{M}$ may have an angle θ against the external magnetic field $\boldsymbol{H}$ applied along the y-axis. We assume that a monochromatic neutron beam incidents onto the sample surface at an angle α_i and scattered at the glancing angle α_f. The incident polarization vector is set either along with, or opposite to $\boldsymbol{H}$ and perpendicular to the scattering plane.

With polarized neutron reflectivity it is possible to measure independently the non spin-flip (NSF) reflectivities R^{++}, R^{--} and the spin-flip (SF) reflectivities R^{+-}, R^{-+}.

The difference of the specularly reflected NSF neutrons is proportional to the Y-component of the magnetic induction B_Y:

$$R^{++} - R^{--} \propto B_Y, \tag{12}$$

whereas the spin-flip reflectivities

$$R^{+-} = R^{-+} \tag{13}$$

are degenerate in the specular direction and are proportional to the square of the X-component of the magnetic induction B_X:

$$R^{+-}, R^{-+} \propto B_X^2. \tag{14}$$

In comparison with vector-MOKE, PNR naturally provides a vector information of the magnetization via the simultaneously occurring NSF and SF reflectivities, without the need of resetting the sample. Furthermore, PNR can distinguish not only between case (a) in Fig. 20, a coherent rotation of magnetization and case (b), nucleation and domain wall motion, but also between case (b) and case (c), domain formation: as any transverse or x component of the magnetization yields spin-flip scattering, while in MOKE experiments the transverse components compensate and do not give rise to a Kerr rotation.

In addition, PNR is depth sensitive and allows the analysis of the magnetization vector even for layers which are deeply buried under covers of other magnetic, or non-magnetic materials. In contrast, MOKE analysis is limited to the near surface region within the skin depth of the particular material, which usually is on the order of 20–30 nm.

For laterally structured thin films with a defined periodicity, such as magnetic stripe arrays, additional reciprocal lattice streaks are generated, which occur in the x-z plane of the reciprocal space at regular spacing on the right and left side to the specular rod. Moreover, stripe arrays have, by design, an easy axis of the shape anisotropy parallel to the stripe direction and a hard axis perpendicular to it. In order to study the reversal mechanism for different orientations of the field with respect to the stripe orientation, the sample needs to be rotated, while keeping the applied field perpendicular to the

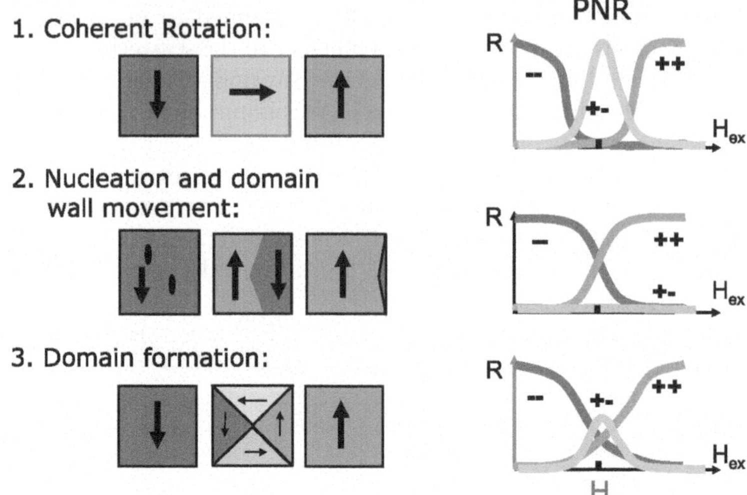

Fig. 20. Sketch of three different possibilities for the magnetization reversal from negative saturation field to positive saturation field. This corresponds to only one branch of a magnetic hysteresis in an ascending field. Panel (a) shows schematically a reversal via a coherent rotation of the magnetization vector together with the respective specularly reflected intensity at a fixed scattering vector. In panel (b) domain nucleation and growth is assumed, and in panel (c) the reversal occurs via domain formation. For more details, see text.

scattering plane and parallel to the polarization axis. Starting from a stripe arrangement perpendicular to the scattering plane (easy axis configuration), rotation of the stripe array implies an effective increase of the lattice period with respect to the scattering plane, resulting in a shrinking separation between the reciprocal lattice rods. In Fig. 19 the scattering geometry is shown together with the definition of the angles, which are used later on for the discussion of examples. The easy axis corresponds to $\chi = 0°$ and the hard axis to $\chi = 90°$.

In the following we discuss the three different reversal mechanisms and how they are expressed in neutron reflectivity measurements. Figure 20 shows schematically in (a) coherent rotation of the magnetization vector, in (b) the domain structure for nucleation and domain wall motion, and (c) reversal via domain formation. The neutron reflectivity is assumed to be taken at one particular value of the scattering vector, which is most sensitive to the magnetization direction, i.e. where the difference $R^{++} - R^{--}$ is largest. Starting from the "down" magnetization in saturation, the difference $R^{--} - R^{++} > 0$, but reverses sign at the coercive field. The cross over with $R^{--} = R^{++}$ occurs at the coercive field H_c. For case (b), $R^{--} = R^{++} > 0$ and at the same time $R^{+-} = R^{-+} = 0$, as no x-component occurs during the magnetization reversal via nucleation and domain wall movement. In contrast, during reversal via coherent rotation the entire magnetization contributes to spin-flip reflection at the coercive field and both non-flip cross sections drop to zero: $R^{--} = R^{++} = 0$. The third case of domain formation is in between the other two cases. At the coercive field spin-flip and non-flip reflection coexist. Reality in most cases is more complex. But the selected reversal types are good guidelines for the discussion of actual examples.

Experimental Results

Magnetization reversal measurements of different magnetic stripe arrays have been carried out using polarized neutron scattering at small angles [3, 38, 39]. The experiments were performed using the polarized neutron reflectometer ADAM at the Institut Laue-Langevin, Grenoble, France [36]. As an illustrative example we will discuss an array consisting of polycrystalline $Co_{0.7}Fe_{0.3}$ stripes with a thickness of 76 nm, a width of 2.4 μm, and a grating period of 3 μm. The anisotropy of the stripes is dominated by the shape anisotropy [39].

In general, polarized neutron scattering data properly analyzed provide a comprehensive picture about the re-magnetization process of lateral magnetic patterns. Basically three different features can be deduced from PNR: Specular reflectivity, Bragg reflections from the lateral periodicity, and off-specular diffuse scattering. Specular and Bragg-PNR measures a signal averaged over a number of domains within the coherence length of the neutron radiation, while magnetization fluctuations on smaller length scales cause off-specular diffuse scattering. All three contributions have been measured and analyzed. In Fig. 21 reflectivity data with spin analysis of the exit beam at two angles of sample rotation $\chi = 0°$ (left column) and $\chi = 75°$ (right column) are reproduced. In saturation we observe the typical ferromagnetic splitting between $R^{++} - R^{--}$ and almost no intensity for the SF reflectivities, $R^{+-} \approx R^{-+} \approx 0$. For $\chi = 0$ (Fig. 21, (a)), in the range $0 \leq H \leq H_c$ the splitting between R^{++} and R^{--} is reduced, and simultaneously spin flip reflection occurs. This is not expected for an easy-axis behavior, for which usually a 180° domain wall propagates through a stripe, causing no SF reflection. Here the SF specular reflection is a first hint to a more complex domain structure in the range $0 \leq H \leq H_c$. For $\chi = 75°$ (Fig. 21 (c)) the SF reflectivity gradually increases and the splitting between R^{++} and R^{--} gradually decreases. At $H = 0$ the co-existence of NSF and SF intensities is due to the remanent state with non-zero x- and y-components of the magnetization vector. For $\chi = 90°$ the y-component should vanish. At $H = H_c$ the NSF-reflectivity reaches zero, while the SF intensity increases towards its maximum. This behavior is typical for any orientation close to the hard axis, which is characterized by a coherent rotation of the magnetization vector.

For a quantitative analysis of the measured neutron intensities, a domain state model was applied and the reflectivities were calculated by use of the super-matrix routine [40, 41]. Within the domain state model it is assumed that each stripe is broken into a set of domains smaller then the lateral

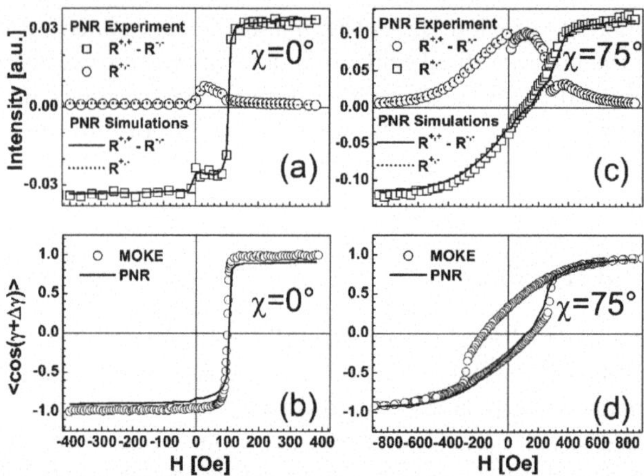

Fig. 21. Polarized neutron reflectivity measurements for tilt angles $\chi = 0°$ (left column) and $\chi = 75°$ (right column). (a) and (c) The difference of the non-flip, $R^{++} - R^{--}$, and the spin flip reflectivities, R^{+-} and R^{-+}, are plotted as a functions of the applied field; (b) and (d) field dependence of the mean projection of the normalized magnetization onto the easy axis as determined from MOKE (circles) and PNR (line) data. From Ref. [38].

projection of the neutron coherence length covering a number of stripes and domains. Magnetization in each domain is tilted away from the direction of the total sample magnetization for an angle $\gamma \pm \Delta\gamma$, where γ is the tilt angle of the mean magnetization of the coherence volume averaged over random deviations $\Delta\gamma$ which are due to individual domains. As a result the total magnetization is reduced by a factor $\langle \cos(\gamma + \Delta\gamma \rangle$ with respect to its value in saturation.

This reduces the spin asymmetry $R^{++} - R^{--}$ during the re-magnetization as seen in Figs. 21(a) and (c), where the experimental data are plotted along with result of the fit to the model. For the sample orientation $\chi = 0$ the total magnetization is directed along the stripes. In Fig. 21(b) the mean value $\langle \cos(\gamma + \Delta\gamma \rangle$ found from the fit is compared with longitudinal MOKE results for the same sample. The agreement is obviously very good, aside from a small deviation between $0 \le H \le H_c$. Also for the sample orientation $\chi = 75°$ the agreement between neutron and MOKE results is excellent.

The agreement between PNR and MOKE results is very satisfactory. At the same time it is clear that (vector-)MOKE is, in comparison, a much faster and a more convenient experimental method than PNR. Therefore, PNR work has to be justified by an information gain which goes beyond the capabilities of vector-MOKE. This is indeed the case when it comes to correlation and fluctuation effects. Then PNR provides detailed and quantitative information about correlation and fluctuation effects, which are difficult or impossible to determine otherwise.

In Fig. 22 we show an intensity map of polarized neutron scattering from the same periodic $Co_{0.7}Fe_{0.3}$ – stripe array as already discussed above. The maps were taken with a position sensitive neutron detector for a sample rotation $\chi = 0°$, a magnetic field of 43 Oe, which is close to the coercive field, and as a function of the glancing incident and exit angles α_i and α_f, respectively. One pattern was recorded with spin-up neutrons (a) and the other pattern was recorded with spin-down neutrons (b). The color code of the intensity map follows a logarithmic intensity scale. The bottom row shows the model calculations simulating the experimental data for spin-up neutrons (c) and spin-down neutrons (d). The specular rod, Bragg reflections of first and second order on either side of the specular ridge, and diffuse scattering at small angles is clearly visible. The off-specular diffuse magnetic scattering shows an asymmetry, which is caused by spin-flip processes due to fluctuations in directions of magnetization around its mean value.

The fluctuations are found to be correlated over distances of at least 20 µm across the stripes. From a quantitative analysis using the distorted wave approximation [40, 41] we infer that the individual domain magnetization directions slightly fluctuate due to small angles $\Delta\gamma$ about the mean value. The mean value itself, determined by the dispersion $\langle \sin^2 \gamma \rangle$, is randomly tilted by almost $\pm 30°$ with respect to the stripes orientation at 43 Oe before switching. Thus polarized specular and off-specular neutron scattering provides a detailed picture of the mean domain magnetization vectors in a magnetic stripe array, including longitudinal and transverse fluctuations about the mean magnetization and correlation effects between magnetic domains across different stripes.

Further Investigations with PNR

Several other studies have been performed on periodic arrays using specular and off-specular polarized neutron reflectivity. The earliest explorative study dates back only a few years [42]. Since then a number of lateral structures have been investigated with neutrons. Among those are the investigation of the reversal mechanism of periodic arrays of rectangular Co bars with a strong shape anisotropy [43, 44, 45], and the reversal of Co/CoO exchange biased stripes [46]. The top layer of a Fe/Cr superlattice was laterally structured into stripes of Fe layers, which then are exchange coupled by a Cr spacer layer to the remaining Fe/Cr superlattice [47]. This structure produces rich diffuse magnetic scattering pattern, which is challenging to analyze. A different approach was taken by the authors of Ref. [48]. Elliptically shaped FeCo bars with constant major axis and alternating size of the minor axis were periodically arranged on a substrate. As the switching field depends on the size of the minor axis, within a certain field range antiferromagnetic alignment of the elliptic bars could established and the specular and off-specular scattering from this array was investigated.

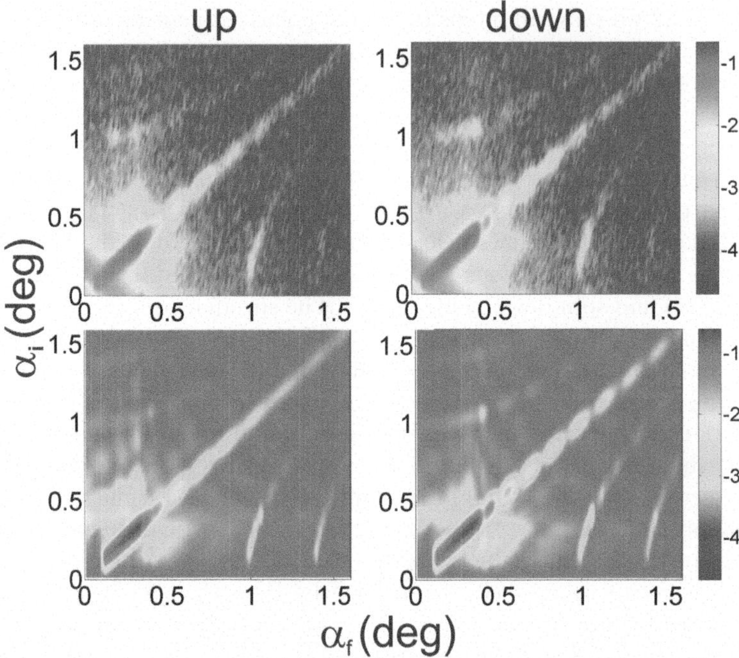

Fig. 22. Top panels: Experimental maps of the polarized neutron intensity on a logarithmic scale from a periodic stripe array measured at a magnetic field of 43 Oe and plotted as a function of the angles of incidence α_i, and the scattering angles α_f. Bottom panels: Calculated intensity maps according to the domain state model for spin up and spin down incident neutron polarization.

4 Summary and Conclusions

We have described four different scattering methods for the characterization of laterally structures magnetic patterns: vector and Bragg magneto-optics, soft x-ray resonant magnetic scattering, and polarized neutron scattering at small angles. They all have in common that they are non-local probes of the constituting magnetic elements and that the magnetic information is gained from a polarization analysis of the scattering beam.

Vector MOKE and polarized neutron scattering are very similar as they both determine the x- and y-components of $\boldsymbol{M}$ allowing a complete analysis of the orientation and magnitude of the magnetization vector as function of applied field. From this analysis, in turn, the magnetization reversal can be characterized as being dominated by nucleation and domain wall motion or by coherent rotation. Nevertheless, both methods, vector-MOKE and polarized neutron scattering, do not measure exactly the same properties. First, MOKE is sensitive to intraband transitions of the ferromagnetic film, while polarized neutron scattering at small angles senses the magnetic induction in the sample plane.

There is another important difference between MOKE and PNR. Vector-MOKE experiments, as presented here, are taken in a specular configuration, while the polarized neutron intensity has been evaluated at Bragg reflections. The difference is one of analyzing the average as compared to correlations. In the language of scattering physics this relates to the self correlation function in case of specular vector-MOKE measurements as compared to pair-correlation in case of polarized neutron scattering. Differences, which have been observed between MOKE and neutron results may be due to this effect. One can circumvent this difference by combining vector-MOKE with Bragg-MOKE, i.e. the MOKE effect at high orders of interference from the stripe pattern. However, this is only possible if the magnetic islands are arranged on a square lattice.

Bragg-MOKE is still in its infancy and needs further theoretical development. Once established, it will be very powerful for the fast and averaging analysis of patterned magnetic media. Correlation effects between the islands as well as the magnetization distribution with the islands can be retrieved from the shape of the hysteresis measured for different orders of interference.

X-ray magnetic resonance scattering combines most of the advantages of the previous methods. Specular reflectivity, off-specular Bragg scattering, and vector magnetometery can be performed. As the soft x-ray wavelengths are an order of magnitude smaller than the wavelenghts of laser beams used in MOKE set-ups, much higher orders of Bragg reflections can be reached. Vice versa, much smaller lattice parameters of the magnetic pattern can be analyzed. However, due to the smaller wavelengths, soft x-rays are less sensitive to the magnetic domain distribution within the islands. Finally, the resonance condition for magnetic scattering guarantees simultaneously element selectivity, which is a feature unique to XRMS.

The strength of neutron scattering is the sensitivity to correlations and fluctuations, the depth resolving vector magnetometery even for layers which are deeply buried, and a magnetic signal, which does not depend on resonance conditions. Hence the cross sections are much easier to interpret and the magnetic moments can be determined on an absolute scale. To summarize, for the analysis of the reversal mechanism of laterally patterned media vector- and Bragg MOKE are very efficient, fast and easy to use methods. Polarized neutron scattering and reflectivity, although more elaborate and costly, offers a more complete analysis of various processes involving correlations and fluctuations. Resonant magnetic x-ray scattering is the first choice, if available. The richness of information, though, has its price, which is the complexity of a quantitative data analysis.

In Table 1 the advantages and disadvantages of the four methods discussed in this contribution are compared.

Table 1. Comparison of the different experimental techniques

	vector-MOKE	Bragg-MOKE	SXRMS	PNR
patterned area	1 mm^2	1 mm^2	1 mm^2	200 mm^2
source	laser	laser	synchrotron	cold neutrons
element selective	no	no	yes	no
depth sensitive	no	no	no	yes
magnetic field restrictions	no	no	no	no
orientation of magnetization vector	in-plane	in-plane	in-plane	in-plane

5 Acknowledgements

This work was supported by *SFB 491* of the Deutsche Forschungsgemeinschaft: "Magnetic Heterostructures: Structure and Electronic Transport". The neutron reflectometer is supported by BMBF under contract O3ZA6BC1 and soft x-ray work facility is supported by the BMBF under contract O3ZA6BC2.

References

1. Jing Shi: Magnetization Reversal in Patterned Magnetic Nanostructures. In: *Ultrathin Magnetic Structures*, vol 4, ed by B. Heinrich and A.C. Bland (Springer, Berlin Heidelberg New York 2004) pp 307–331
2. C. Daboo, R. J. Hicken, E. Gu, M. Gester, S. J. Gray, D. E. P. Eley, E. Ahmad, J. A. C. Bland, R. Poessl, J. N.Chapman, Phys. Rev. B, **51**, 15964 (1995).
3. K. Theis-Bröhl, T. Schmitte, V. Leiner, Zabel, K. Rott, H. Brückl, and J. McCord, Phys. Rev. B **67**, 184415 (2003)
4. M. Grimsditch, P. Vavassori, J. Phys.: Condens. Matter **16**, 275 (2004).
5. T. Schmitte, K. Westerholt, H. Zabel, J. Appl. Phys. **92** (8), 4524 (2002)
6. P. Vavassori, V. Methloshko, R.M. Osgood III, M. Grimsditch, U.Welp, G. Crabtree, W. Fan, S.R.J. Brueck, B. Ilic, P.J. Hesketh, Phys. Rev. B **59**, 6337 (1999).
7. O. Geoffroy, D. Givord, Y. Otani, B. Pannetier, A. Santos, M. Schlenker, J. Magn. Magn. Mater. **121**, 516 (1993).
8. I. Guedes, N.J. Zaluzec, M. Grimsditch, V. Methloshko, P. Vavassori, B. Ilic, P. Neuzil, R. Kumar, Phys. Rev. B 62, 11719 (2000).
9. M. Grimsditch, P. Vavassori, V. Novosad, V. Metlushko, H. Shima, Y. Otani, and K. Fukamichi, Phys. Rev. B **65**, 172419 (2002).

10. P. Vavassori, N. Zaluzec, V. Metlushko, V. Novosad, B. Ilic, and M. Grimsditch, Phys. Rev. B **69**, 214404 (2004).
11. M. J. Donahue, D. G. Porter, Technical Report No. NISTIR 6376, National Institute of Standards and Technology (1999).
12. A. Westphalen, K. Theis-Bröhl, H. Zabel, K. Rott, H. Brückl, J. Mag. Mag. Mater., article in press (2005).
13. A. Westphalen, T. Schmitte, K. Westerholt, H. Zabel, J. Appl. Phys. **97**, 073909 (2005).
14. I. Guedes, M. Grimsditch, V. Metlushko, P. Vavassori, R. Camley, B. Ilic, P. Neuzil, R. Kumar, Phys. Rev. B **67**, 024428 (2003).
15. P. Vavassoria, M. Grimsditch, V. Novosad, V. Metlushko, B. Ilic, J. Appl. Phys. **93**, 7900 (2003).
16. V. Novosad, M. Grimsditch, J. Darrouzet, J. Pearson, S. D. Bader, V. Metlushko, K. Guslienko, Y. Otani, H. Shima, and K. Fukamichi, Appl. Phys. Lett. **82**, 3716 (2003).
17. F. Radu, A. Westphalen, K. Theis-Bröhl, H. Zabel, J. Phys.: Condens. Matter **18** L29 (2006).
18. A. Westphalen, A. Remhof, K. Theis-Bröhl, H. Zabel, in preparation.
19. T. Schmitte, K. Theis-Brohl, V. Leiner, H. Zabel, S. Kirsch, A. Carl, J. Phys.: Cond. Mat. **14** 7527 (2002).
20. D.A. Allwood, G. Xiong, M.D. Cooke, R.P. Cowburn, J. of Physics D: Appl. Phys. **36** 2175 (2003).
21. J.L.Erskine, E.A.Stern, Phys. Rev. B **12** 5016 (1975).
22. G. Schütz, W. Wagner, W. Wilhelm, P. Kienle, Phys. Rev. Lett. **58** 737 (1987).
23. J. Grabis, A. Bergmann, A. Nefedov, K. Westerholt, H. Zabel, Phys. Rev. B **72** 024438 (2005).
24. J. Geissler, E. Goering, M. Justen, F. Weigand, G. Sch¨tz, J. Langer, D. Schmitz, H. Maletta, and R. Mattheis, Phys. Rev. B **65** 020405(R) (2002).
25. U. Hillebrecht, Science **284** 2099 (1999).
26. H. Zabel and K. Theis-Bröhl, J. Phys.: Condens. Matter **15** 505 (2003).
27. G. van der Laan, K. Chesnel, M. Belakhovsky, A. Marty, F. Livet, S.P. Collins, E. Dudzik, A. Haznar, J.P. Attané, Superlattices and Microstructures **34** 107 (2003).
28. K. Chesnel, M. Belakhovsky, S. Landis, J.C. Toussaint, S.P. Collins, G. van der Laan, E. Dudzik, and S.S. Dhesi, Phys. Rev. B **66** 024435 (2002).
29. Carlo Spezzani, Mauro Fabrizioli, Patrizio Candeloro, Enzo Di Fabrizio, Giancarlo Panaccione, and Maurizio Sacchi, Phys. Rev. B **69** 224412 (2004).
30. J.P. Hannon, G.T. Trammel, M. Blume, D. Gibbs, Phys. Rev. Lett. **61** 1245 (1988); Phys. Rev. Lett. **62** 2644 (1989).
31. G. van der Laan, B.T. Thole, Phys. Rev. B **43** 13401 (1991).
32. A. Remhof, C. Bircan, A. Westphalen, J. Grabis, A. Nefedov, H. Zabel, Superlattices and Microstructures **37** 353 (2005).
33. J. Grabis, A. Nevedov, H. Zabel, Rev. Scientific Instr. **74** 4048 (2003).
34. T. Last, S. Hacia, M. Wahle, S. F. Fischer, U. Kunze, J. Appl. Phys. **96** 6706 (2004).
35. H.A. Dürr, E. Dudzik, S.S. Dhesi, J.B. Goedkoop, G. van der Laan, M. Belakhovsky, C. Mocuta, A. Marty, Y. Samson, Science **284** 2166 (1999).
36. A. Schreyer, R. Siebrecht, U. Englisch, U. Pietsch, and H. Zabel Physica B **248** 349 (1998).

37. C.F. Majkrzak, Physica B **156** & **157** 619 (1989).
38. K. Theis-Bröhl, H. Zabel, J. McCord, B. Toperverg, Physica B **356** 14 (2005).
39. K. Theis-Bröhl, V. Leiner, A. Westphalen, H. Zabel, J. McCord, K. Rott, H. Brckl, B. P. Toperverg, Phys. Rev. B **71**, 020403 (2005).
40. B.P. Toperverg, Physica B **297** 160 (2001).
41. B.P. Toperverg, Appl. Phys. A **74** 1560 (2002).
42. B.P. Toperverg, G.P. Felcher, V.V. Metlushko, V. Leiner, R. Siebrecht, O. Nikonov, Physica B **283**, 149 (2000).
43. K. Temst, M.J. Van Bael, H. Fritzsche, Appl. Phys. AMater **74** 1538 (2002).
44. K. Temst, M.J. Van Bael, J. Swerts, D. Buntinx C. Van Haesendonck, Y. Bruynseraede, H. Fritzsche, R. Jonckheere, J. Vac. Sci. Technol. B **21**, 2043 (2003).
45. H. Fritzsche, M.J. Van Bael, K. Temst, Langmuir **19** 7789 (2003).
46. K. Temst, M.J. Van Bael, J. Swerts, H. Loosvelt, E. Popova, D. Buntinx, J. Bekaert, C. Van Haesendonck, Y. Bruynseraede, R. Jonckheere, H. Fritzsche, Superlattices and Microstructures **34**, 87 (2003).
47. N. Ziegenhagena, U. Rücker, E. Kentzinger, R. Lehmann, A. van der Hart, B. Toperverg, Th. Brückel, Physica B **335**, 50 (2003).
48. K. Theis-Bröhl, et al. to be published.

Finite-Temperature Simulations for Magnetic Nanostructures

M.A. Novotny[1,2], D.T. Robb[1,2,3], S.M. Stinnett[1,2,4], G. Brown[3,5], and P.A. Rikvold[3,6,7]

[1] Center for Computational Sciences, Mississippi State University, Mississippi State, MS, 39762-9627 USA novotny@erc.msstate.edu
[2] Department of Physics and Astronomy, Mississippi State University, Mississippi State, MS, 39762-5167 USA
[3] School of Computational Science, Florida State University, Tallahassee, FL, 32306-4120 USA browngrg@scs.fsu.edu, rikvold@scs.fsu.edu, robb@scs.fsu.edu
[4] Department of Physics, McNeese State University, Box 93140, Lake Charles, LA 70609 sstinnett@mcneese.edu
[5] Center for Computational Sciences, Oak Ridge National Laboratory, Oak Ridge, TN, 37831-6164 USA
[6] Center for Materials Research and Technology, Florida State University, Tallahassee, FL, 32306-4350 USA
[7] Department of Physics, Florida State University, Tallahassee, FL 32306-4350 USA

Summary. We examine different models and methods for studying finite-temperature magnetic hysteresis in nanoparticles and ultra-thin films. This includes micromagnetic results for the hysteresis of a single magnetic nanoparticle which is misaligned with respect to the magnetic field. We present results from both a representation of the particle as a one-dimensional array of magnetic rotors, and from full micromagnetic simulations. The results are compared with the Stoner-Wohlfarth model. Results of kinetic Monte Carlo simulations of ultra-thin films are also presented. In addition, we discuss other topics of current interest in the modeling of magnetic hysteresis in nanostructures, including kinetic Monte Carlo simulations of dynamic phase transitions and First-Order Reversal Curves.

1 Introduction

Although hysteresis in fine magnetic particles has been intensively studied for many years, there is currently significant interest in reexamining our understanding of this phenomenon. Partly, this interest is driven by the potential application of hysteresis in nanostructures to new technologies such as Magnetic Random Access Memory (MRAM) and ultra-high-density magnetic recording. For the past several years, the areal density of hard drives has been doubling every 18 months, and is rapidly approaching the limits of conventional longitudinal recording technology. At the same time, data rates in these drives have increased significantly, with the serial interface standard at the time of writing providing a peak data transfer rate of 2.4 Gb/s [1]. This has led the magnetic recording industry to look at new recording paradigms such as patterned media and self-assembled arrays of nanostructures. In fact, the first laptop computer incorporating a hard drive based on perpendicular recording technology was recently introduced [2]. It is crucial, then, to understand the complex process of hysteresis in these systems.

At the same time, recent advances in computational ability, both in terms of new algorithms and available computer resources, allow for numerical studies never before possible. Plumer and van Ek [3], for instance, have studied the effects of anisotropy distributions in perpendicular media using a micromagnetic model. Their results (Fig. 1) show how anisotropy distributions tend to reduce the squareness of the loop and, therefore, the signal to noise ratio (SNR). Gao et al. have recently carried out similar studies of tilted perpendicular media [4] and polycrystalline media [5]. Another important effect which can be better understood through simulations is Barkhausen noise. This effect also decreases SNR, particularly in new thin-film media with soft underlayers. Dahmen, Sethna, and coworkers used a random-field Ising model to examine the origins of Barkhausen noise and have been able to relate it to avalanches and disorder-induced critical behavior [6, 7, 8]. These results illustrate two of the ways simulations can be used to help understand both fundamental processes in hysteresis and their applications to new technology.

Here we present an overview of several common approaches to studying hysteresis in magnetic nanostructures. We then present results of large-scale computer simulations of hysteresis in single iron nanoparticles when the magnetic field is misoriented with respect to the long (easy) axis of the elongated particles. We also examine other recent advances in the study of magnetic hysteresis, such as kinetic Monte Carlo simulations of dynamic phase transitions and First-Order Reversal Curves.

2 Models and Methods

2.1 Coherent Rotation

Given a single-domain particle with uniaxial anisotropy, it is possible to find the metastable and stable energy positions of the magnetization when a magnetic field is applied at an angle to the easy axis. It is assumed that the magnetization can be represented by a single vector $\boldsymbol{M}$, with constant amplitude, M_S. The energy density of the system is then

$$E = K \sin^2 \vartheta - M_S H \cos(\phi - \vartheta) , \tag{1}$$

where K is the uniaxial anisotropy constant, H is the magnetic field applied at an angle ϕ to the easy axis, and ϑ is the angle the magnetization makes with the easy axis. Stoner and Wohlfarth showed that for coherent reversal of the magnetization, the spinodal curve beyond which the metastable energy minimum disappears and switching occurs is given by [9],

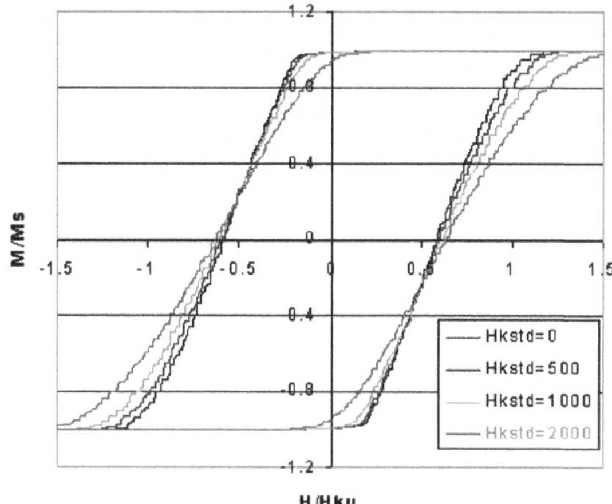

Fig. 1. The effect of anisotropy distributions on hysteresis in perpendicular media. A Gaussian distribution of anisotropy fields is used with a mean value of $H_{ku} = 7000\,\mathrm{Oe}$. In the bottom left corner, the curves correspond to standard deviations in H_K of (from left to right) 2000 Oe, 1000 Oe, 500 Oe, and 0 Oe. The saturation magnetization is $350\,\mathrm{emu/cm}^3$. Data courtesy of M. Plumer [3]

$$h_{AX}^{2/3} + h_{AY}^{2/3} = 1 , \tag{2}$$

where h_{AX} and h_{AY} are the respective components of the magnetic field normalized by the anisotropy field $H_K = 2K/M_S$, along the easy and hard axes. Equation (2) is the well-known equation of a hypocycloid of four cusps, also known as an astroid.

2.2 Micromagnetics

For systems in which the spins are not aligned and/or the field is changing too rapidly for the magnetization to reach its quasi-static value, it is usually necessary to use a non-perturbative technique such as micromagnetics to describe the reversal process. The basic approach is to divide the system into a coarse-grained set of sites. Each site is associated with a position $\boldsymbol{r}_i$, and its magnetization is represented by a single magnetization vector $\boldsymbol{M}(\boldsymbol{r}_i)$, whose norm is the saturation magnetization M_S, corresponding to the bulk material (a valid assumption for temperatures well below the Curie temperature [10]). The time evolution of each spin is given by the Landau–Lifshitz–Gilbert (LLG) equation [11, 12, 13],

$$\frac{\mathrm{d}\boldsymbol{M}(\boldsymbol{r}_i)}{\mathrm{d}t} = \gamma_0 \left(\boldsymbol{M}(\boldsymbol{r}_i) \times \left[\boldsymbol{H}(\boldsymbol{r}_i) - \frac{\alpha}{\gamma_0 M_S} \left(\frac{\mathrm{d}\boldsymbol{M}(\boldsymbol{r}_i)}{\mathrm{d}t} \right) \right] \right) , \tag{3}$$

where $\boldsymbol{H}(\boldsymbol{r}_i)$ is the total local field at the i-th site, γ_0 is the gyromagnetic ratio ($1.76 \times 10^7\,\mathrm{rad/Oe\,s}$), and α is a dimensionless damping parameter which determines the rate of energy dissipation in the system. The first term represents the precession of each spin around the local field, while the second term drives the magnetization to align with the field. The LLG equation can easily be rewritten in a form more convenient for numerical integration [11, 14]

$$\frac{\mathrm{d}\boldsymbol{M}(\boldsymbol{r}_i)}{\mathrm{d}t} = \frac{\gamma_0}{1+\alpha^2}\left(\boldsymbol{M}(\boldsymbol{r}_i) \times \left[\boldsymbol{H}(\boldsymbol{r}_i) - \frac{\alpha}{M_S}\left(\boldsymbol{M}(\boldsymbol{r}_i) \times \boldsymbol{H}(\boldsymbol{r}_i)\right)\right]\right) . \quad (4)$$

For the sign of the undamped precession term, we follow the convention of Brown [11].

The total local field, $\boldsymbol{H}(\boldsymbol{r}_i)$, controls the dynamics and contains all of the interactions between each site and the rest of the system; it is defined by

$$\boldsymbol{H}(\boldsymbol{r}_i) = -\frac{\partial E_i}{\partial \boldsymbol{M}(\boldsymbol{r}_i)} . \quad (5)$$

Here, E_i is the free energy of the i-th site and the operator $\partial/\partial\boldsymbol{M}(\boldsymbol{r}_i) = (\partial/\partial M_x(\boldsymbol{r}_i))\hat{x} + (\partial/\partial M_y(\boldsymbol{r}_i))\hat{y} + (\partial/\partial M_z(\boldsymbol{r}_i))\hat{z}$. The different terms that contribute to $\boldsymbol{H}(\boldsymbol{r}_i)$ combine via linear superposition,

$$\boldsymbol{H}(\boldsymbol{r}_i) = \boldsymbol{H}_Z(\boldsymbol{r}_i) + \boldsymbol{H}_e(\boldsymbol{r}_i) + \boldsymbol{H}_D(\boldsymbol{r}_i) + \boldsymbol{H}_a(\boldsymbol{r}_i) + \boldsymbol{H}_n(\boldsymbol{r}_i) . \quad (6)$$

Here, $\boldsymbol{H}_Z(\boldsymbol{r}_i)$ is the externally applied field (Zeeman term), $\boldsymbol{H}_e(\boldsymbol{r}_i)$ is due to exchange interactions, $\boldsymbol{H}_D(\boldsymbol{r}_i)$ is the dipole field, $\boldsymbol{H}_a(\boldsymbol{r}_i)$ is the anisotropy field (in our simulations taken to be zero), and $\boldsymbol{H}_n(\boldsymbol{r}_i)$ is a random field representing the effects of thermal noise.

The exchange contribution to the local field represents local variations between the alignment of $\boldsymbol{M}(\boldsymbol{r}_i)$ and neighboring sites and can be represented by $l_e^2\nabla^2\boldsymbol{M}(\boldsymbol{r}_i)$ [14]. In our simulations, this is implemented by

$$\boldsymbol{H}_e(\boldsymbol{r}_i) = \left(\frac{l_e}{\Delta r}\right)^2 \left(-n_i\boldsymbol{M}(\boldsymbol{r}_i) + \sum_{|\boldsymbol{d}|=\Delta r} \boldsymbol{M}(\boldsymbol{r}_i + \boldsymbol{d})\right) , \quad (7)$$

where the summation is over the nearest neighbors of $\boldsymbol{r}_i$, n_i is the number of neighbors of site i, and the term $n_i\boldsymbol{M}(\boldsymbol{r}_i)$ is included so that $\boldsymbol{H}_e = 0$ when all of the spins are aligned. The exchange length, l_e, is defined in terms of the exchange energy [15], $E_e = -\left(l_e^2/2\right)\int d\boldsymbol{r}\boldsymbol{M}\cdot\nabla^2\boldsymbol{M}$, in a *continuous* system. For our discrete system of finite-sized cells, this means the magnetization can be viewed as rotating continuously from the center of one cell to the center of each neighboring cell along the line joining the two.

At non-zero temperatures, thermal fluctuations contribute a term to the local field in the form of a stochastic field $\boldsymbol{H}_n(\boldsymbol{r}_i)$, which is assumed to fluctuate independently for each spin. The fluctuations are assumed to be Gaussian, with zero mean and (co)variance given by the fluctuation-dissipation theorem [11, 13],

$$\langle H_{n\mu}(\boldsymbol{r}_i,t)H_{n\mu'}(\boldsymbol{r}'_i,t')\rangle = \frac{2\alpha k_B T}{\gamma_0 M_S V}\delta(t-t')\delta_{\mu,\mu'}\delta_{i,i'}\ , \tag{8}$$

where $H_{n\mu}(\boldsymbol{r}_i)$ indicates one of the Cartesian coordinates of $\boldsymbol{H}_n(\boldsymbol{r}_i)$. Here, T is the absolute temperature, k_B is Boltzmann's constant, $V = (\Delta r)^3$ is the discretization volume of the numerical integration, and $\delta_{r,r'}$ is the Kronecker delta representing the orthogonality of the Cartesian components. Although this result was derived for an isolated particle, recent work by Chubykalo, et al. indicates that this result will hold for interacting systems as well [16]. In this paper, we present micromagnetic results for two different models. The first model is of a nanoparticle with dimensions 5.2 nm × 5.2 nm × 88.4 nm. The cross-sectional dimensions are small enough ($\approx 2\,l_e$) that the assumption is made that the only significant inhomogeneities occur along the long axis (z-direction) [12, 17]. The particles of this model are therefore discretized into a linear chain of 17 spins along the long axis of the particle. We will call this model the stack-of-spins model.

In this simple model, the local field due to dipole-dipole interactions is calculated as [15, 17]

$$\boldsymbol{H}_D(\boldsymbol{r}_i) = (\Delta r)^3 \sum_{j\neq i} \frac{3\hat{\boldsymbol{r}}_{ij}(\hat{\boldsymbol{r}}_{ij}\cdot\boldsymbol{M}(\boldsymbol{r}_j)) - \boldsymbol{M}(\boldsymbol{r}_j)}{\boldsymbol{r}_{ij}^3}\ , \tag{9}$$

where $\boldsymbol{r}_{ij}$ is the displacement vector from the center of cube i to the center of cube j, and $\hat{\boldsymbol{r}}_{ij}$ is the corresponding unit vector. The volume factor $(\Delta r)^3$ results from integration over the constant magnetization density in each cell. The second model, which we will refer to as the full micromagnetic model, simulates a single nanoparticle with dimensions 9 nm × 9 nm × 150 nm. The dimensions were chosen to correspond to arrays of nanoparticles fabricated by Wirth, et al. [18]. In this model, the system is discretized into 4949 sites (7 × 7 × 101) on the computational lattice. The size of the system makes calculation of dipole interactions in the conventional manner (as done for the stack-of-spins model) computationally impractical. It is therefore necessary to use a more advanced algorithm to make the simulation tractable.

The two most popular choices are the traditional Fast Fourier Transform (FFT) and the Fast Multipole Method (FMM) [19]. Here, we used the Fast Multipole Method, the exact implementation of which is discussed elsewhere [13], because it has several advantages over the FFT. The biggest difference is that the FMM makes no assumptions about the shape of the underlying lattice, while the FFT assumes a cubic lattice with periodic boundary conditions. The consequence of this is that numerical models of systems without periodic boundary conditions which use the FFT require empty space around the system so that the boundary conditions do not affect the calculation. The FMM requires no such "padding". Furthermore, the FFT requires $O(N\ln N)$ operations to calculate the magnetic scalar potential (from which the dipole field is calculated). The FMM algorithm, while it has a larger computational overhead, requires only $O(N)$ operations for the same calculation.

This means that, while the FFT is a good choice for small cubic lattices, the FMM is better for large, incomplete, or irregular lattices. The public-domain psi-Mag toolset now provides a flexible implementation of the FMM designed for use on high performance, parallel computers [20].

Material properties in both models were chosen to correspond to bulk Fe. The saturation magnetization is 1700 emu/cm^3 (kA/m) and $l_e = 2.6$ nm. We take the damping parameter $\alpha = 0.1$ to correspond to the underdamped behavior usually assumed to be present in nanoscale magnets. Although this value is approximately an order of magnitude larger than the value obtained experimentally using Ferromagnetic Resonance (FMR), it has been noted that the FMR value is for small deviations of the magnetization from equilibrium and is not representative of the large deviations which occur during reversal [21]. In general, care should be taken in establishing an appropriate damping parameter to use in simulating a particular magnetic nanostructure, as also illustrated in other recent micromagnetic studies [22, 23, 24].

It is worth noting that, even in systems which reverse coherently at high speed, deviations from the quasi-static Stoner–Wohlfarth (SW) model will be expected. He, et al. [25] showed that for square-pulse fields with fast rise times (< 10 ns) and small values of the damping constant (< 0.2), the shape of the astroid changes. The result is that the minimum switching field is reduced below the SW limit of 0.5 H_K, and the angular dependence is no longer symmetric around 45°. For the frequencies and damping parameter used here, the deviations from the SW model are small (< 5 percent difference in the switching fields) and may be neglected.

2.3 Monte Carlo Simulations: Kinetic Ising and Heisenberg Models

A second approach to modeling the dynamics of magnetic systems involves Monte Carlo techniques, which have been applied to a wide variety of systems since their introduction by Metropolis, et al. [26]. As described above, the micromagnetics approach uses a stochastic (i.e. random number-based) variable to introduce random fluctuations into an otherwise deterministic system. In contrast, Monte Carlo simulations are fully stochastic and proceed by considering possible transitions between states of the system and executing these transitions with a probability which depends on the system's energy and temperature.

The static Monte Carlo algorithm consists of a repeated three-step process. First, choose a (pseudo-)random number. The random numbers chosen may be uniformly distributed, or they may be chosen based on a particular probability distribution that depends on the specific simulation to be performed. Second, choose a trial move from the current state to a new state. Third, accept or reject the trial move depending on the random number and some acceptance rule consistent with the problem under consideration.

Consider the ferromagnetic Ising model on a regular lattice with periodic boundary conditions. Each site on the lattice has a spin which can align either parallel or anti-parallel to the applied field and takes on values of $S_i = \pm 1$ accordingly. The energy of the Ising lattice is then

$$E = -J \sum_{\langle i,j \rangle} S_i S_j - H(t) \sum_i S_i \,, \tag{10}$$

where the exchange constant $J > 0$ is in units of energy, and $H(t)$ is the externally applied, time-dependent magnetic field (which in Monte Carlo simulations is customarily given in units of energy, thus absorbing the magnetic moment per site, μ). The first sum in (10) is over nearest-neighbor pairs, while the second sum is over all spins on the lattice.

The static Monte Carlo procedure described above allows the calculation of equilibrium quantities such as the internal energy, susceptibility, specific heat, and magnetization. In order for the lattice to explore each of its possible states with probabilities corresponding to the equilibrium thermal distribution, the acceptance rule chosen must satisfy the condition of detailed balance [27]. Two common choices are the Metropolis [26] and Glauber [28] acceptance rules. Note that near the critical point, computation with these simple acceptance rules slows down dramatically, and it is therefore useful to use more advanced algorithms (such as cluster algorithms [27]) to calculate equilibrium quantities.

In equilibrium calculations, no physical interpretation is ascribed to the intermediate spin flips. If, instead, we consider the individual spin flips as representing physical fluctuations due to the interactions between the spins and a heat bath, then the underlying transitions model the actual dynamics of the system and acquire a physical significance. This application of Monte Carlo simulations is known as kinetic Monte Carlo. The random nature of the events due to the interaction of spins dictates that the spin to attempt to flip must be chosen at random. In this paper, we use the Glauber [28] acceptance rule, according to which each attempted spin flip is accepted with probability

$$W = \frac{\exp\left(-\beta \Delta E_i\right)}{1 + \exp\left(-\beta \Delta E_i\right)} \,. \tag{11}$$

Here, ΔE_i is the change in energy that results if the proposed flip of the i-th spin is accepted, and $\beta = (k_B T)^{-1}$. With a uniformly distributed random number, $r \in [0, 1]$, a randomly chosen spin is flipped if $r \leq W$. Each potential spin flip is considered a Monte Carlo step. The basic time step of the Monte Carlo process is measured in Monte Carlo Steps per Site (MCSS). This time is related to the algorithm and in general is only approximately proportional to the physical time of the system. Recently, however, progress has been made in connecting analytically the MC simulation time to the simulation time of the Langevin-based micromagnetic techniques discussed above, for which there is a clear relationship to physical time [29, 30, 31, 32].

The Glauber dynamic of (11) can be derived from a quantum spin-$\frac{1}{2}$ Hamiltonian coupled to a fermionic heat bath [33]. Recently, other dynamics have been derived from coupling a quantum spin-$\frac{1}{2}$ system to a phonon heat bath [34]. Note that in kinetic Monte Carlo calculations, algorithms (such as the cluster algorithm) that change the underlying dynamic cannot be used. However, advanced algorithms that achieve very large speedups while remaining true to the underlying dynamics are possible [35]. It has recently been shown that physically relevant functional forms for W can lead to dramatically different values of dynamical quantities such as lifetimes of metastable states [36].

It is important to realize that the Monte Carlo techniques can be applied to other systems as well. Unlike the Ising model, the Heisenberg model allows the spins to assume any angle with respect to neighboring spins and the applied field. The energy of a regular lattice of Heisenberg spins with periodic boundary conditions is

$$E = -J \sum_{\langle i,j \rangle} (S_{ix}S_{jx} + S_{iy}S_{jy} + S_{iz}S_{jz}) - H \sum_{i} S_i \cos(\theta_i) , \qquad (12)$$

where S_{ix}, S_{iy}, and S_{iz} are the Cartesian coordinates of the vector spin $\mathbf{S}_i$ (with magnitude $S_i = 1$), and θ_i is the angle between the applied field, H, and the i-th spin. As in the Ising model, the first sum is taken over nearest neighbors and represents the exchange interactions, while the second sum is taken over all spins in the system, and represents the interactions of the spins with an externally applied magnetic field (Zeeman energy).

The dynamic consists of randomly choosing a spin to update, randomly choosing a new spin direction (either uniformly distributed over the sphere or over a cone near the current spin direction), and using a Metropolis or a heat-bath rate to decide whether to effect a transition to the new spin direction. The rate depends on the energy difference between the spin configurations as in, for example, (11).

In kinetic Monte Carlo, it is possible to implement the algorithm in a rejection-free manner, so that every algorithmic step performs an update. In this case, each algorithmic step in general advances the system by a different amount of time. For models, such as the Ising model, with discrete state spaces this is called the n-fold way algorithm [37]. It is possible to make a precise connection between these the n-fold way and the standard implementation of kinetic Monte Carlo [35]. Recently such rejection-free methods have been implemented for models with continuous state spaces, such as the Heisenberg model [38], and the efficiency of rejection-free methods in various systems has been studied [39].

3 Results of Micromagnetic Simulations

In this section we summarize recent simulation results for magnetization reversal in iron nanopillars [40], and further evaluate these results in light of additional experimental data on such reversal.

Figure 2a shows hysteresis loops at $T = 100\,\mathrm{K}$ for the full micromagnetic model with the field misaligned at 0°, 45°, and 90° to the long axis of the particle. The loops were calculated using a sinusoidal field with a period of 25 ns, which started at a maximum value of 10,000 Oe (800 kA/m). In all the loops in this section, the reported magnetization is the component along the long axis (z-axis) of the particle. Simulations for the full micromagnetic model were performed over one half of the period and the results reflected to give the full hysteresis loop.

Consider the case with the field and particle aligned (0°). Initially, the large magnetic field tends to align the spins with the easy axis. As the field is decreased, the spins relax, and the magnetization decreases by approximately 2%. Eventually, reversal initiates at the ends as previously reported [13].

At 45° misalignment between the particle and the field, the magnetization is initially pulled away from the long (easy) axis by the large magnetic field. As the field is swept toward zero, the magnetization relaxes until it essentially reaches a maximum value of approximately $0.91M_S$ at zero applied field. Thermal fluctuations along the length of the particle prevent the magnetization from reaching saturation. As in the case of 0°, reversal again begins by nucleation at the ends of the particle, with the growth of these nucleated regions leading to the reversal of the particle. Figure 3a shows the z-component of the magnetization at selected times during the reversal process for the 45° hysteresis loop of Fig. 2a. It is important to note that the particles do not have a uniform magnetization during the reversal process, even though they are single-domain particles.

For 90° misalignment, the reversal mechanism is quite different. The hysteresis loop in Fig. 2a shows that the magnetization is essentially perpendicular to the easy direction until the field reaches a particular value. As the field is decreased further, the magnetization relaxes toward the easy axis. Since nothing breaks the up/down symmetry of the system when the applied field has no component along the easy axis, the relaxed magnetization can be directed toward either the positive or negative z-axis. Figure 3b shows the z-component of the magnetization for the 90° misalignment at selected times during the hysteresis loop of the full micromagnetic model. For this case, the nucleation occurs along the entire length of the particle, except at the ends. The large demagnetizing fields present at the ends (involved in nucleation at smaller angles) retard relaxation along the easy axis.

The hysteresis loops for the stack-of-spins model, shown in Fig. 2b, are qualitatively similar to those of Fig. 2a. Loops at 0°, 75°, and 90° misalignment are shown. There are important differences between the two models, however. First, without lateral resolution of the magnetization across the

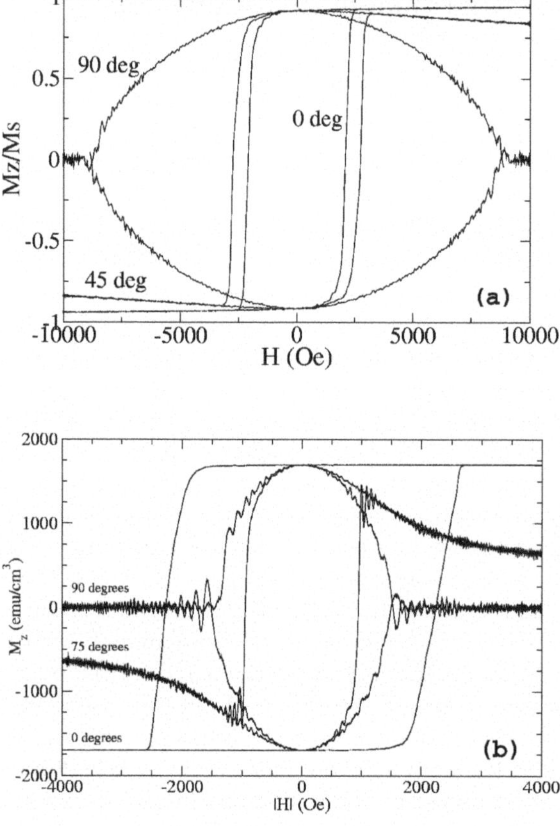

Fig. 2. Hysteresis loops for the full micromagnetic model and the stack-of-spins model. In (**a**), the full micromagnetic model with 0°, 45°, and 90° misalignment, at $T = 100\,\mathrm{K}$ with a sinusoidal field of period 25 ns and amplitude 10 kOe. In (**b**), the stack-of-spins model with 0°, 75°, and 90° misalignment, at $T = 10\,\mathrm{K}$ with a sinusoidal field of period 200 ns and amplitude 5 kOe

cross-section, these particles exhibit ringing due to the precessional dynamics. Evidently, the precession of individual moments in the full micromagnetic model does not lead to precession of the end-cap moment; possibly the spin waves rapidly damp out the gyromagnetic motion.

A second, and more prominent, difference between the models is observed in the angular dependence of the switching field, H_{sw}, shown in Fig. 4. Here, H_{sw} is defined as the applied field at which M_z is reduced to 0. The stack-of-spins model (circles) shows a shape qualitatively similar to what is expected from Stoner–Wohlfarth (SW) theory, with a minimum H_{sw} near 45°. The dashed curve is the SW theory with $H_K = 1600\,\mathrm{Oe}$ (for comparison reasons H_K was chosen to be much smaller than the 10^4 Oe expected for these parti-

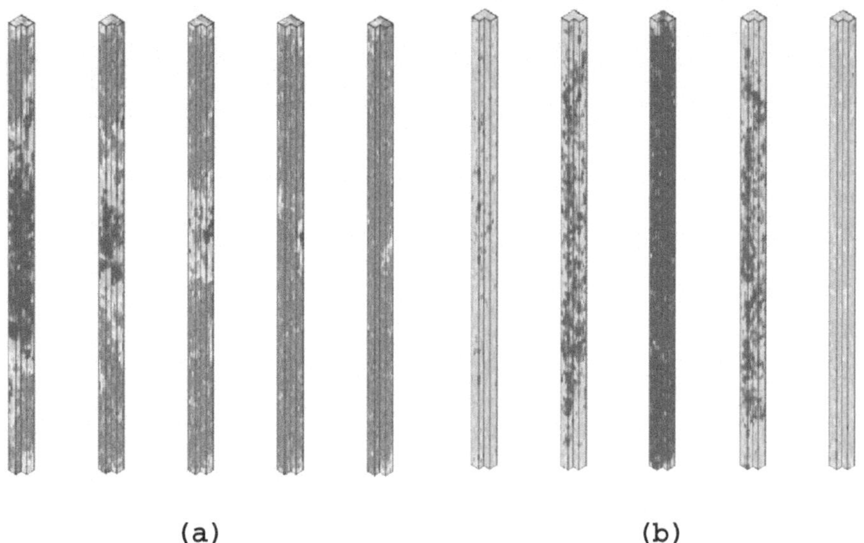

Fig. 3. The z-component of the magnetization in the full micromagnetic model for the (**a**) 45° and (**b**) 90° hysteresis loops of Fig. 1a. In (**a**), the snapshots were taken at times (left to right) $t = 7.350$, 7.375, 7.400, 7.425, and 7.450 ns. In (**b**), the snapshots were taken at times $t = 0.00$, 3.000, 4.000, 6.250, and 8.500 ns

cles assuming SW behavior). The full micromagnetic model (diamonds), on the other hand, has its minimum H_{sw} at 0°, and H_{sw} increases as the misalignment angle is increased. Figure 5 shows the angular dependence of the switching field for the full micromagnetic model for periods of 15, 25, 50, and 100 ns and maximum applied field of 5 kOe at $T = 0$ K, and for periods of 15 ns and 25 ns with a maximum applied field of 10 kOe at $T = 100$ K. At 0 K, the general trend is for longer periods to reduce the switching field. However, at 15°, the 100 ns loop is observed to switch at a lower field than either the 50 or 25 ns loops. Similarly, at 30°, the 50 ns loop switches at a lower field than the 25 ns loop. One reason for this may be resonance in the switching fields for these angles and periods. At $T = 100$ K, the 25 ns loops switch at a lower field than the 15 ns loops for all angles. At 90°, where thermal fluctuations are most prominent, the field at which relaxation occurs is independent of the period within the accuracy of the simulation.

The increase of H_{sw} with the misalignment angle in the micromagnetic simulation is consistent with recent experimental observations of Fe nanopillars [18, 41, 42, 43, 44]. However, the most recent experiment [44] shows that a nanopillar with lateral dimension $d \sim 5.2$ nm, which our formulation suggests should show a dependence of H_{sw} on misalignment angle similar to the stack-of-spins model (i.e. like coherent rotation), actually exhibits the increasing dependence found in the full micromagnetic model. In addition, a nanopillar with lateral dimension $d \sim 10$–15 nm showed evidence

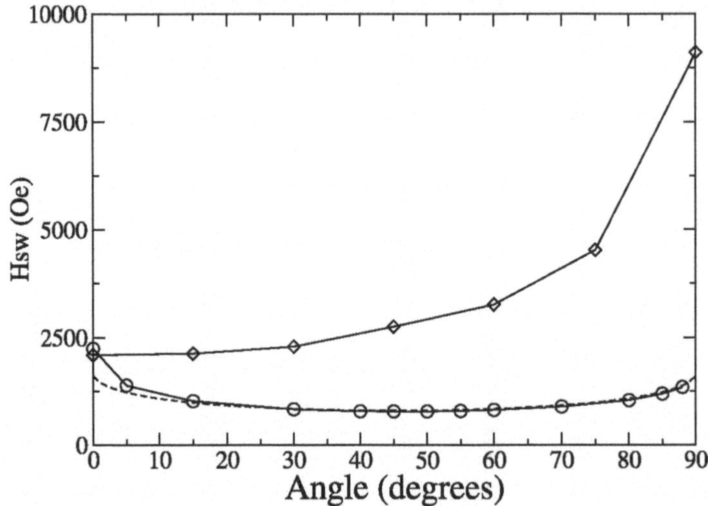

Fig. 4. Angular dependence of the switching field for three models of magnetization reversal. The full micromagnetic model (*diamonds*) at $T = 100\,\mathrm{K}$ shows a distinctly different behavior from both the stack-of-spins model at $T = 20\,\mathrm{K}$ (*circles*) and the Stoner–Wohlfarth model (*dashed line*)

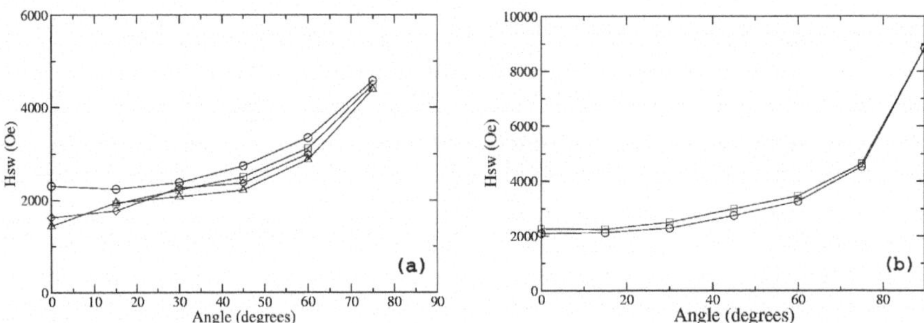

Fig. 5. Angular dependence of the switching field for the full micromagnetic model at (**a**) 0 K and (**b**) 100 K. At 0 K, the LLG equation is completely deterministic, while at 100 K, it includes random fluctuations through a stochastic thermal field. In (**a**), the applied field periods are 15 ns (*circles*), 25 ns (*squares*), 50 ns (*diamonds*), and 100 ns (*triangles*), and the field is sinusoidal with amplitude 5 kOe. In (**b**), the applied field periods are 15 ns (*squares*) and 25 ns (*circles*), and the field is sinusoidal with amplitude 10 kOe

of a multi-domain remanence state. As noted in [44], imperfections of the nanopillar structure appear to contribute to localized nucleation processes down to smaller than expected lateral dimensions, and probably also provide the pinning sites causing the multi-domain remanence state. This illustrates the importance of coordinating experimental and simulation results in the micromagnetic approach. Further improvement of the predictions of the micromagnetic approach will likely have to incorporate such structural imperfections.

4 Recent Results for the 2D Kinetic Ising Model

Monte Carlo simulation of the Ising model, as well as other magnetic systems, continues to be an active field of research. Here, we present three recent results that are of interest in understanding the process of magnetization reversal in ultra-thin films.

4.1 Dynamic Phase Transitions

When the half-period $t_{1/2}$ of the applied field is longer than the characteristic switching time in a constant field, $\langle\tau(H_0)\rangle$, where H_0 is the amplitude of the oscillating field, the magnetization can follow the changing field, resulting in standard hysteresis loops, such as those shown in Fig. 1 and in Fig. 6a. However, when $t_{1/2} \ll \langle\tau(H_0)\rangle$, the magnetization cannot follow the field, but rather oscillates around one or the other of its zero-field stable values. This breaking of the symmetry of the hysteresis loop is associated with a dynamic phase transition (DPT) located at an intermediate value of the half-period. In terms of the dimensionless half-period, $\Theta = t_{1/2}/\langle\tau(H_0)\rangle$, the transition is located at $\Theta \approx 1$. The dynamic order parameter for this transition is the period-averaged magnetization,

$$Q_n = \frac{1}{2t_{1/2}} \int_{(n-1)(2t_{1/2})}^{n(2t_{1/2})} m(t)\mathrm{d}t \; . \tag{13}$$

In Fig. 6, we show hysteresis loops for the two-dimensional kinetic Ising model using Glauber dynamics for the dynamically disordered phase with $\Theta \gg \Theta_c$ and the dynamically ordered phase with $\Theta \ll \Theta_c$. Time series of Q_n for $\Theta \gg \Theta_c$, $\Theta \approx \Theta_c$, and $\Theta \ll \Theta_c$ are shown in Fig. 7.

The DPT was first discovered in numerical solutions of a mean-field model of a ferromagnet in an oscillating field [45, 46]. It has since been intensively studied in mean-field models [47, 48, 49], kinetic Ising models [50, 51, 52, 53, 54, 55], the kinetic spherical model [56], and anisotropic XY [57, 58] and Heisenberg [59, 60] models. There have also been indications of its presence in experimental studies of hysteresis in ultra-thin films of Cu on Co(001) [61, 62]. From a theoretical point of view, its most interesting feature is that this *far-from-equilibrium* phase transition is a genuine continuous (second-order) phase transition that belongs to the *same* universality class as the *equilibrium* phase transition in the Ising model in zero field [53, 54, 55, 63]. Unequivocal experimental verification of this interesting non-equilibrium phase transition

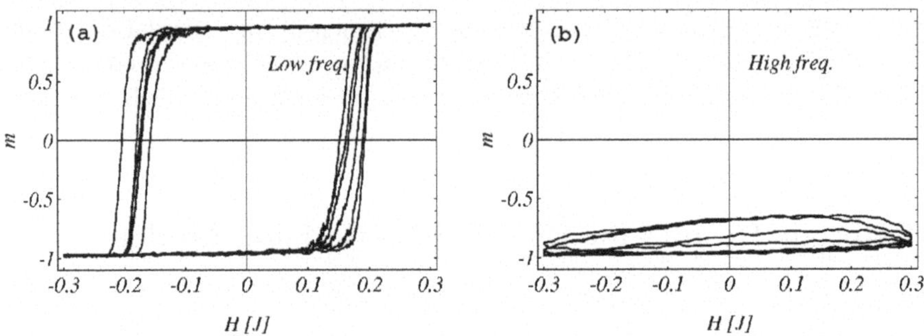

Fig. 6. Simulated hysteresis loops for a kinetic Ising model (**a**) in the dynamically disordered phase for $\Theta \gg \Theta_c$ and (**b**) in the dynamically ordered phase for $\Theta \ll \Theta_c$. Data courtesy of S.W. Sides

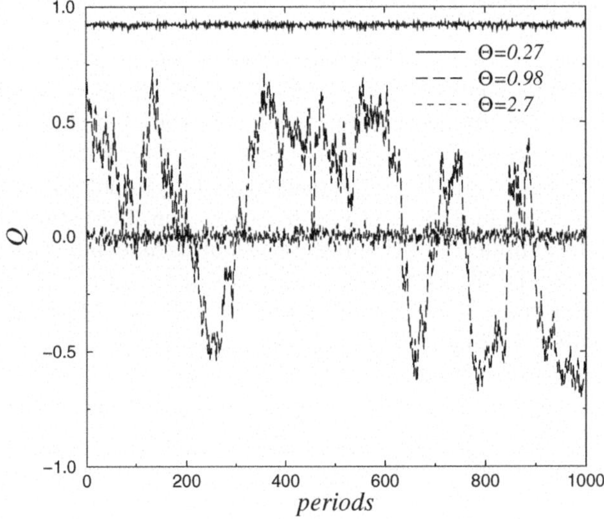

Fig. 7. Time series of the dynamic order parameter Q_n in the dynamically ordered phase (curve near $+1$, $\Theta = 0.27$), near the transition (curve fluctuating wildly about zero, $\Theta = 0.98$), and in the dynamically disordered phase (curve that remains close to zero, $\Theta = 2.7$). After [54]

is highly desirable and, given that new high-density magnetic recording media will require shorter reversal periods, may be relevant to the design of magnetic storage devices.

4.2 Hysteresis Loop Area and Stochastic Resonance

The values of Q measured for the stack-of-spins model described above appear to be consistent with the existence of a DPT, although no detailed

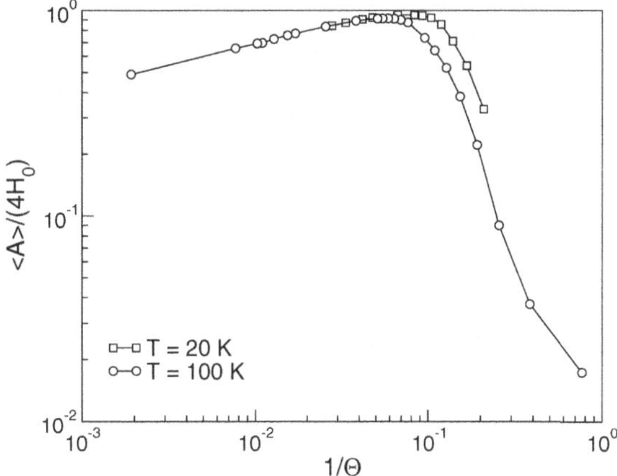

Fig. 8. Average hysteresis-loop area, $\langle A \rangle$, vs scaled frequency, $1/\Theta$ for the stack-of-spins micromagnetic model. The same behavior is seen in two-dimensional Ising models that switch by a single-droplet mechanism, and the maximum is associated with stochastic energy resonance

analysis has yet been made [64]. At lower frequencies, another interesting behavior is seen in both the stack-of-spins and kinetic Ising models [64, 65]. The normalized average hysteresis-loop area,

$$\langle A \rangle = -\frac{1}{4M_S H_0} \oint M(H)\,\mathrm{d}H \,, \tag{14}$$

is a measure of the average energy dissipation per period and is therefore a very important quantity. It is shown vs scaled frequency, $1/\Theta$, in Fig. 8 for the stack-of-spins model at $T = 100\,\mathrm{K}$ and $T = 20\,\mathrm{K}$. At extremely low frequencies, the magnetization switches at very small values of H, so that $\langle A \rangle \approx 0$. At high frequencies, the switching rarely completes because the system is metastable for only a very short time interval. Therefore, M is nearly constant and again $\langle A \rangle \approx 0$. A maximum in $\langle A \rangle$ occurs at intermediate frequencies $1/\Theta \approx 0.1$. For studies of hysteresis in a kinetic Ising model which switches by a single-droplet mechanism, this maximum was found to correspond to stochastic energy resonance [65]. This phenomenon has been studied further in the kinetic Ising model [55, 66, 67], and also recently investigated in models of superparamagnetic nanoparticles [68] and Preisach systems [69].

4.3 First-Order Reversal Curves

The First-Order Reversal Curve (FORC) technique was developed by Pike, et al. [70] in order to extract more information from magnetic samples than is represented by, for example, the coercive field or the remanent magnetization.

The FORC method has since been applied to a wide variety of systems, including several relevant to magnetic nanostructures [73]–[80]. In addition, progress has been made in understanding the role of reversible magnetization in the FORC method [81] and in improving the efficiency of its computational use [82]. Here, we illustrate the basic approach with an application to the kinetic Ising model.

The FORC technique involves decreasing the applied field from a positive saturating field, H_0, to a series of progressively more negative return fields, H_r, and recording the normalized magnetization, $m = M/M_S$, as the field is increased from each of these return fields back to the positive saturating field. This process results in a family of first-order reversal curves, $m(H_r, H)$, where H represents the applied magnetic field as it is increased from H_r back to H_0. Since the first-order reversal curves (FORCs) are determined by the type of reversal that has taken place before reaching H_r, the full family of FORCs should contain useful information about the mechanisms of reversal.

We can use the FORC method to better understand the process of hysteresis in the two-dimensional ferromagnetic kinetic Ising model on a square lattice, choosing the Glauber acceptance rule to produce the dynamic of the system with the energy given by (10). While most FORC studies have been done on systems with strong disorder, we focus here on the square-lattice Ising model without disorder. Our simulations were performed at a temperature of $T = 0.8\ T_c$ which, given that $k_B T_c \approx 2.269J$ for the two-dimensional square-lattice Ising model, corresponds to $k_B T \approx 1.815J$. It has been found [83] that the switching of a fully magnetized lattice for these parameters occurs through single-droplet nucleation for fields up to $|H| \approx 0.35$, by multi-droplet nucleation for fields $|H| \approx 0.35$–0.9, and by strong-field (single-spin) reversal for fields $|H| > 0.9$. Since the process of switching is also influenced by the lattice size for finite lattices, these values serve only as guidelines. Here, we are mainly concerned with the multi-droplet regime, and so choose $H_0 = 0.55$.

We performed MC simulations to calculate the characteristic switching time τ (for switching from $m = 1.0$ to $m < -0.8$) in a field of magnitude $H_0 = -0.55$, finding $\tau \approx 100$ MCSS for a 128×128 lattice. We therefore chose a field period of $P = 1000$ MCSS, corresponding to a dimensionless half-period $\Theta = \frac{P/2}{\tau} \approx 5$. The form of the field is taken as a sawtooth, piecewise linear function

$$H(t) = H_0 \left(\frac{4|t - P/2|}{P} - 1 \right) . \tag{15}$$

Figure 9a shows the results of the simulation on a 128×128 lattice for dimensionless half-periods of $\Theta = 5$, 10, and 25. The simulations were performed in parallel with 100 independent realizations distributed over 20 processors using the 48-bit linear congruential random number generator included with the SPRNG 2.0 package [84].

As the lattice just completes its reversal during the full hysteresis loop, we expect that the family of FORCs will reflect much of the dynamics that

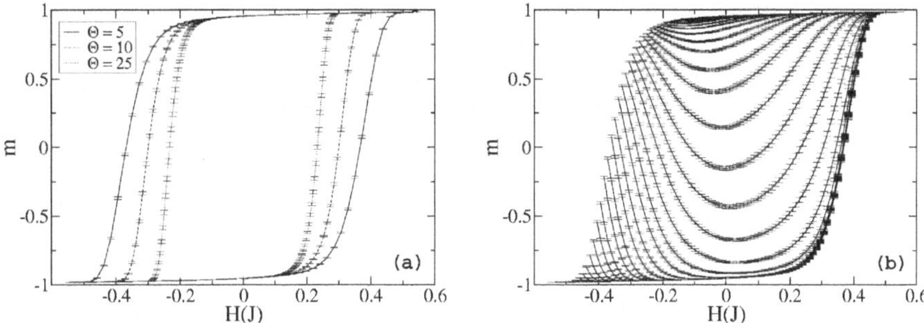

Fig. 9. (**a**) Hysteresis loops for $\Theta = 5$ (solid line), 10 (dashed line), and 25 (dotted line) on a 128×128 Ising lattice. (**b**) Family of FORCs for the same lattice with $\Theta = 5$

are occurring during the reversal. To investigate this, we divided the interval from $H = [-0.55, 0.55]$ into 100 equal intervals. We began the first FORC at a return field of $H_r = 0.0$, and recorded the magnetization at H_a values corresponding to the endpoints of the 100 intervals. (Thus, for the first FORC, we took 51 values of the magnetization.) We then took a series of FORCs for H_r values at the interval endpoints from $H = 0.0$ to $H = -0.55$, producing a total of 51 FORCs. For each curve, we averaged over 100 realizations of the MC simulation, a technique commonly used to find the thermally averaged behavior of a system. The resulting family of FORCs is shown in Fig. 9b. An animation of the reversal process for the FORCs shows that the reversal does proceed by the nucleation, growth, and shrinkage of multiple droplets (i.e., areas of reversed magnetization).

In a recent article [85], we have continued this investigation of the kinetic Ising model using the family of FORCs, as well as the FORC distribution, which can be derived from the FORCs as described in [70]. The analysis yielded insights into the limits of application of the Kolmogorov-Johnson-Mehl-Avrami (KJMA) model of phase transformation [86, 87, 88] to the kinetic Ising system. In general, the FORC method appears to be quite sensitive to details of the magnetization reversal process, and with some thought can be helpful in developing insights into the construction of useful models.

5 Conclusion

Information storage devices utilizing magnetic nanostructures have become a technologically important part of our society. As demands for information storage increase, the size of the nanostructures must be decreased. At the same time, it becomes important to read and write the information to these devices (i.e. reverse the magnetization) faster. The understanding of hysteresis in the magnetic nanostructures is therefore important to the continued growth of the information-storage industry. At the same time, the growth of computational resources has provided researchers with an invaluable tool with which to better understand these systems.

In this overview, various common models and methods for simulating hysteresis in magnetic nanostructures have been presented along with results illustrating some of the properties of these systems. Micromagnetic simulations are accomplished by integration of the Landau-Lifshitz-Gilbert (LLG) equation. The LLG equation, despite being both classical and phenomenological in origin, nevertheless provides good insight into the magnetization dynamics at nanosecond time scales, provided the system is sufficiently finely discretized. Our simulations on single Fe nanopillars show that the switching field (i.e. the field required to reduce M_Z to 0) increases continuously as the angle between the z-axis and the applied field direction is increased, consistent with experiment. Reversal in these pillars is shown to nucleate at the endcaps and proceed by domain growth toward the center of the particle. The exception to this is the case of the applied field perpendicular to the long axis of the pillar, in which nucleation of reversal occurs along the whole length of the particle.

Unfortunately, limitations on computer resources prevent extension of micromagnetic simulations beyond timescales of a few tens of nanoseconds. For timescales where the transition time for an individual spin to relax from the metastable to the stable state is much shorter than the time scale of interest, individual spin reversals occur with a probability which is related to the Boltzmann factor. The dynamics of the system can then be modeled using kinetic Monte Carlo techniques with either the Ising or Heisenberg models. Here, we have shown three interesting applications of kinetic Monte Carlo simulations of a 2-D Ising model to understanding hysteresis: dynamic phase transitions, stochastic resonance in the hysteresis loop area, and First-Order Reversal Curves (FORCs). These illustrate only a few of the ways simulations of magnetic nanostructures may help give new insight into this important class of materials for ultra-high-density data storage.

Acknowledgments

This work was supported in part by NSF grants No. DMR-0120310 and DMR-0444051, and by the DOE Office of Science through the Computational Materials Science Network of BES-DMSE.

References

1. http://en.wikipedia.org/wiki/Serial_ATA
2. eWeek, Jan. 17, 2006 (formerly PC Week magazine, now online only at http://www.eweek.com)

3. M. Plumer, J. van Ek: *Micromagnetic System Modeling for Perpendicular Recording*, presented at the Micromagnetics and Magnetic Recording Workshop held at the Center for Materials for Information Technology, the University of Alabama (November 6, 2003) (unpublished)
4. K.Z. Gao, H.N. Bertram: IEEE Trans. Magn. **39**, 704 (2003)
5. K.Z. Gao, J. Fernandez-de-Castro, H.N. Bertram: IEEE Trans. Magn. **41**, 4236 (2005)
6. O. Perković, K. Dahmen, J.P. Sethna: Phys. Rev. Lett. **75**, 4528 (1995)
7. K. Dahmen, J.P. Sethna: Phys. Rev. B **53**, 14872 (1996)
8. J.H. Carpenter, K. Dahmen, J.P. Sethna, G. Friedman, S. Loverde, A. Vanderveld: J. Appl. Phys. **89**, 6799 (2001)
9. E.C. Stoner, E.P. Wohlfarth: Phil. Trans. Roy. Soc. **A240**, 599 (1948)
10. D.A. Garanin: Phys. Rev. B **55**, 3050 (1997)
11. W. Brown: *Micromagnetics* (Wiley, 1963)
12. U. Nowak: Thermally Activated Reversal in Magnetic Nanostructures. In: *Annual Reviews of Computational Physics IX*, ed. by D. Stauffer (World Scientific, Singapore 2001) pp. 105–152
13. G. Brown, M.A. Novotny, and P.A. Rikvold: Phys. Rev. B **64**, 134432 (2001)
14. A. Aharoni: *Introduction to the Theory of Ferromagnetism* (Clarendon, Oxford 1996)
15. A.S. Arrot, B. Heinrich, D.S. Bloomberg: IEEE Trans. Magn. **MAG-10**, 950 (1974)
16. O. Chubykalo, R. Smirnov-Rueda, J.M. Gonzalez, M.A. Wongsam, R.W. Chantrell, U. Nowak: J. Magn. Magn. Mater. **266**, 28 (2003)
17. E. Boerner, H.N. Bertram: IEEE Trans. Magn. **33**, 3052 (1997)
18. S. Wirth, M. Field, D.D. Awschalom, S. von Molnár: Phys. Rev. B **57**, R14028 (1998)
19. L.F. Greengard: *The Rapid Evaluation of Potential Fields in Particle Systems* (MIT Press, Cambridge 1988)
20. G. Brown, T.C. Schulthess, D.M. Apalkov, P.B. Visscher: IEEE Trans. Magn. **40**, 2146 (2004)
21. W.D. Doyle, S. Stinnett, C. Dawson, L.He: J. Magn. Soc. Jpn. **22**, 91 (1998)
22. J. Fidler, T. Schrefl, W. Scholz, D. Suess, V.D. Tsiantos, R. Dittrich, M. Kirschner: Physica B **343**, 200 (2004)
23. R. Hertel, J. Kirschner: J. Magn. Magn. Mater. **270**, 364 (2004)
24. E.D. Boerner, K.Z. Gao, R.W. Chantrell: IEEE Trans. Magn. **40**, 2371 (2004)
25. L. He, W.D. Doyle, H. Fujiwara: IEEE Trans. Magn. **30**, 4086 (1994)
26. N. Metropolis, A.W. Rosenbluth, M.N. Rosenbluth, A.H. Teller, E. Teller: J. Chem. Phys. **21**, 1087 (1953)
27. D. Landau and K. Binder: *A Guide to Monte Carlo Simulations in Statistical Physics* (Cambridge University Press, Cambridge 2000)
28. R.J. Glauber: J. Math. Phys. **4**, 294 (1963)
29. U. Nowak, R.W. Chantrell, E.C. Kennedy: Phys. Rev. Lett. **84**, 163 (2000)
30. O. Chubykalo, U. Nowak, R. Smirnov-Rueda, M.A. Wongsam: Phys. Rev. B **67**, 064422 (2003)
31. X.Z. Cheng, M.B.A. Jalil, H.K. Lee, Y. Okabe: Phys. Rev. B **72**, 094420 (2005)
32. X.Z. Cheng, M.B.A. Jalil, H.K. Lee, Y. Okabe: http://www.arxiv.org/cond-mat/0602011 (2006) (to appear in Phys. Rev. Lett.)
33. P. A. Martin: J. Stat. Phys. **16**, 149 (1977)

34. K. Park, M.A. Novotny, P.A. Rikvold: Phys. Rev. E **66**, 056101 (2002)
35. M.A. Novotny: A Tutorial on Advanced Dynamic Monte Carlo Methods for systems with Discrete State Spaces. In: *Annual Reviews of Computational Physics IX*, ed. by D. Stauffer (World Scientific, Singapore 2001) pp. 153–210 (also available at http://www.arxiv.org/cond-mat/0109182)
36. K. Park, P.A. Rikvold, G.M. Buendía, M.A. Novotny: Phys. Rev. Lett., **92** 015701 (2004); G.M. Buendía, P.A. Rikvold, K. Park, M.A. Novotny: J. Chem. Phys. **121**, 4193 (2004); G.M. Buendía, P.A. Rikvold, M. Kolesik: Phys. Rev. B **73**, in press (2006)
37. A.B. Bortz, M.H. Kalos, and J.L. Lebowitz: J. Comput. Phys. **17**, 10 (1975)
38. J.D. Muñoz, M.A. Novotny, S.J. Mitchell: Phys. Rev. E **67**, 026101 (2003)
39. H. Watanabe, S. Yukawa, M.A. Novotny, N. Ito: http://www.arxiv.org/cond-mat/0508652 (2005) (submitted to Phys. Rev. Lett.)
40. G. Brown, S.M. Stinnett, M.A. Novotny, P.A. Rikvold: J. Appl. Phys. **95**, 6666 (2004)
41. S. Wirth, S. von Molnár: J. Appl. Phys. **85**, 5249 (1999)
42. Y. Li, P. Xiong, S. von Molnár, S. Wirth, Y. Ohno, H. Ohno: Appl. Phys. Lett. **80**, 4644 (2002)
43. Y. Li, P. Xiong, S. von Molnár, Y. Ohno, H. Ohno: J. Appl. Phys. **93**, 7912 (2003)
44. Y. Li, P. Xiong, S. von Molnár, Y. Ohno, H. Ohno: Phys. Rev. B **71**, 214425 (2005)
45. T. Tomé, M.J. de Oliveira: Phys. Rev. A **41**, 4251(1990)
46. J.F.F. Mendes, J.S. Lage: J. Stat. Phys. **64**, 653 (1991)
47. P. Jung, G. Gray, R. Ray, P. Mandel: Phys. Rev. Lett. **65**, 1873 (1991)
48. M.F. Zimmer: Phys. Rev. E **47**, 3950 (1993)
49. E.Z. Meilikhov: JETP Letters **79**, 620 (2004)
50. W.S. Lo, R.A. Pelcovits: Phys. Rev. A **42**, 7471 (1990)
51. M. Acharyya, B. Chakrabarti: Phys. Rev. B **52**, 6550 (1995)
52. M. Acharyya: Phys. Rev. E **56**, 2407 (1997)
53. S.W. Sides, P.A. Rikvold, M.A. Novotny: Phys. Rev. Lett. **81**, 834 (1998); Phys. Rev. E **59**, 2710 (1999)
54. G. Korniss, C.J. White, P.A. Rikvold, M.A. Novotny: Phys. Rev. E **63**, 016120 (2000)
55. G. Korniss, P.A. Rikvold, M.A. Novotny: Phys. Rev. E **66**, 056127 (2002)
56. M. Paessens, M. Henkel: J. Phys. A **36**, 8983 (2003)
57. T. Yasui, H. Tutu, M. Yamamoto, H. Fujisaka: Phys. Rev. E **66**, 036123 (2002); erratum: *ibid.* **67**, 019901(E) (2003)
58. N. Fujiwara, H. Tutu, H. Fujisaka: Phys. Rev. E **70**, 066132 (2004)
59. H. Jang, M.J. Grimson, C.K. Hall: Phys. Rev. B **67**, 094411 (2003)
60. Z.G. Huang, F.M. Zhang, Z.G. Chen, Y.W. Du: Eur. Phys. J. B **44**, 423 (2005)
61. Q. Jiang, H.-N. Yang, G.C. Wang: Phys. Rev. B **52**, 14911 (1995)
62. Q. Jiang, H.-N. Yang, G.C. Wang: J. Appl. Phys. **79**, 5122 (1996)
63. H. Fujisaka, H. Tutu, P.A. Rikvold: Phys. Rev. E **63**, 016109 (2001); erratum: *ibid.* **63**, 059903(E) (2001)
64. G. Brown, M.A. Novotny, P.A. Rikvold: Physica B **306**, 117 (2001)
65. S.W. Sides, P.A. Rikvold, M.A. Novotny: Phys. Rev. E **57**, 6512 (1998)
66. M. Acharyya: Phys. Rev. E **59**, 218 (1999)
67. B.J. Kim, P. Minnhagen, H.J. Kim, M.Y. Choi, G.S. Jeon: Europhys. Lett. **56**, 333 (2001)

68. Y.L. Raikher, V.I. Stepanov, R. Perzynski: Physica B **343**, 262 (2004)
69. R.N. Mantegna, B. Spagnolo, L. Testa, M. Trapanese: J. Appl. Phys. **97**, 10E519 (2005)
70. C.R. Pike, A. Roberts, K. Verosub: J. Appl. Phys. **85**, 6660 (1999)
71. C.R. Pike, A.P. Roberts, M.J. Dekkers, K. Verosub: Phys. Earth Planet. Inter. **126**, 11 (2001)
72. A. Roberts, C.R. Pike, K. Verosub: J. Geophys. Res. **105**, 28461 (2001)
73. C.R. Pike, A. Fernandez: J. Appl. Phys. **85**, 6668 (1999)
74. C. Carvallo, A.R. Muxworthy, D.J. Dunlop, W. Williams: Earth Planet. Sci. Lett. **213**, 375 (2003)
75. P.G. Bercoff, M.I. Oliva, E. Borclone, H.R. Bertorello: Physica B **320**, 291 (2002)
76. M.I. Oliva, H.R. Bertorello, P.G. Bercoff: J. Alloys Compd. **354**, 203 (2004)
77. L. Spinu, A. Stancu, C. Radu, F. Li, J.B. Wiley: IEEE Trans. Magn. **40**, 2116 (2004)
78. J.E. Davies, O. Hellwig, E.E. Fullerton, G. Denbeaux, J.B. Kortright, K. Liu: Phys. Rev. B **70**, 224434 (2004)
79. C.R. Pike, C.A. Ross, R.T. Scalettar, G.T. Zimanyi: Phys. Rev. B **71**, 133407 (2005)
80. J.E. Davies, O. Hellwig, E.E. Fullerton, J.S. Jiang, S.D. Bader, G.T. Zimanyi, K. Liu: Appl. Phys. Lett. **86**, 262503 (2005)
81. C.R. Pike: Phys. Rev. B **68**, 104424 (2003)
82. D. Heslop, A.R. Muxworthy: J. Magn. Magn. Mater. **288**, 155 (2005)
83. P.A. Rikvold, H. Tomita, S. Miyashita, S.W. Sides: Phys. Rev. E **49**, 5080 (1994)
84. The SPRNG random number generator is maintained at Florida State University and may be downloaded from http://sprng.cs.fsu.edu/
85. D.T. Robb, M.A. Novotny, P.A. Rikvold: J. Appl. Phys. **97**, 10E510 (2005)
86. A.N. Kolmogorov: Bull. Acad. Sci. USSR, Phys. Ser. **1**, 355 (1937).
87. W.A. Johnson, R.F. Mehl: Trans. Am. Inst. Mining and Metallurgical Engineers **135**, 416 (1939)
88. M. Avrami: J. Chem. Phys. **7**, 1103 (1939); **8**, 212 (1940); **9**, 177 (1941)

Part III

Functional Elements of MRAMs

The Influence of Substrate Treatment on the Growth Morphology and Magnetic Anisotropy of Epitaxial CrO_2 Films

Guo-Xing Miao[1,2], Gang Xiao[2], and Arunava Gupta[1]

[1] Center for Materials for Information Technology, University of Alabama, Tuscaloosa, Alabama 35487

[2] Physics Department, Brown University, Providence, Rhode Island 02912

Summary. Epitaxial CrO_2 thin films deposited on HF-cleaned TiO_2 (100) substrates exhibit very strong strain anisotropy, while those grown without the HF treatment step are essentially strain-free and display bulk-like magnetic properties. The HF treatment enhances the surface smoothness of the TiO_2 (100) substrates thus leading to the growth of epitaxially strained CrO_2 films. The magnetic easy axis of these films changes orientation with thickness, switching from the in-plane c-axis direction for thick films to the b-axis direction for thinner films (< 50 nm). Similarly, over a thickness range, a change of the easy axis direction is also observed with lowering temperature. Ion-beam irradiation of the substrate surface prior to growth also results in the growth of strained CrO_2 films, although the amount of strain is less than that observed for HF-treated substrates. The magnetic properties as a function of thickness have also been studied for as-deposited "thick" CrO_2 films that are slowly chemically etched down in thickness. Unlike as-grown thin films below 50 nm thickness that have the easy axis along the b direction, the chemically etched down CrO_2 thin films of equivalent thickness retain their easy axis alignment along the c-direction, but display a significantly enhanced coercivity. The observed differences in the switching behavior between the as-deposited films and those that are chemically etched can be qualitatively attributed to changes in the strain relaxation mechanism.

1 Introduction

There has been much interest in recent years in studying the magnetotransport properties of chromium dioxide (CrO_2) because of its potential application in the emerging field of spintronics. Band structure calculations have shown that CrO_2 is a half-metallic material, i.e. it contains a gap in the minority spin channel at the Fermi level and no gap in the majority spin channel, resulting in complete spin polarization at the Fermi level [1, 2, 3]. Indeed, point-contact Andreev reflection measurements have provided conclusive evidence for its half-metallicity, with a spin polarization value as high as 98.4% being reported [4, 5]. The high degree of spin polarization, together with a reasonably high Curie temperature of 390 K, makes CrO_2 an attractive candidate for use in magnetoelectronic devices such as magnetic tunneling junctions and spin valves. For these applications it is very important to understand and control the switching behavior of the magnetic layers in the structure.

In this chapter we present results on the magnetic anisotropy of epitaxial CrO_2 films grown on (100)-TiO_2 substrates as a function of different surface treatment conditions. The substrate cleaning procedure used prior to CrO_2 film growth plays an essential role in determining its growth morphology and the resulting magnetic switching characteristics. This results primarily from the effect of strain that influences the magnetic anisotropy of the film. This strain effect can be exploited for the growth of multilayer structures with different switching fields for the individual layers, or even with the layers having mutually perpendicular magnetic anisotropy directions. In contrast to the standard approaches of modifying the growth conditions and film thickness to achieve variations in the coercive field (HC), the use of surface treatment methods as described here offers several advantages, particularly for epitaxial structures.

2 Experimental

CrO_2 has a tetragonal rutile crystal structure (a = 4.421 Å, c = 2.916 Å) that is isostructural with the rutile phase of TiO_2 (a = 4.594 Å, c = 2.958 Å). For the growth of CrO_2 films on (100)-oriented single crystal TiO_2 substrate, the lattice mismatch is anisotropic, being –3.77 and –1.42 along the [001] and [010] directions, respectively. We are able to grow epitaxial CrO_2 films on (100)-oriented TiO_2 substrates by chemical vapor deposition (CVD) using chromium trioxide (CrO_3) as a precursor. Details of the CVD growth have been reported previously [6]. In brief, oxygen is used as a carrier gas in a two-zone furnace to transport the precursor from the source region to the reaction zone where it decomposes selectively on the substrate to form CrO_2. The films are grown at a substrate temperature of about 400°C, with the source temperature maintained at 260°C, and an oxygen flow rate of 100 sccm. The films have been extensively characterized using atomic force microscopy (AFM) and x-ray diffraction. In order to determine the unit cell lattice parameters as a function of film thickness we have carried out x-ray diffraction measurements at angles both around the (200) normal Bragg peak and also the off-axis (110) and (101) peaks. Magnetic measurements have been performed using a Quantum Design superconducting quantum interference device (SQUID) magnetometer and a vibrating sample magnetometer (VSM). While the SQUID has been used primarily for the temperature-dependent studies, the angular dependence of the hysteresis loops at room temperature is measured using the VSM. For the magnetic measurements, it is essential that the back side and edges of the substrates are carefully cleaned to avoid unwanted contributions to the magnetic signal from deposited material on these surfaces.

3 Results and Discussion

3.1 Film Structure and Morphology

The growth morphology of the CrO_2 films is critically dependent on the TiO_2 substrate cleaning procedure utilized prior to deposition. Films grown on as-purchased polished (100)-TiO_2 substrates that are cleaned with organic solvents (acetone and isopropanol) and then rinsed in distilled water are relatively rough and have a columnar morphology with little residual strain. The strain effect is much more pronounced in films grown on substrates that, in addition to the organic clean described above, are briefly treated with dilute HF and then water rinsed and dried prior to deposition. Hereafter, we refer to the CrO_2 films grown on TiO_2 substrates that do not or do undergo the additional HF treatment as films of type A and B, respectively. The normalized lattice parameters of CrO_2 films as a function of thickness, as determined from normal and off-axis x-ray measurements, are plotted in Fig. 1 for both the type A and type B samples. It is clear from the data that the type A films are nearly strain-free. On the other hand, the type B films exhibit a strong influence of strain that varies with thickness, with the lattice parameters gradually approaching the bulk values for the thicker layers. The surface morphology of the substrates after cleaning, and that of the films, has been characterized using AFM. Figure 2(a) and (b) shows AFM images of the TiO_2 substrate surface after only organic clean and with additional HF treatment, respectively. While the RMS roughness for both surfaces is comparable, the HF treated surface is much better ordered locally as evidenced by the appearance of atomic steps ($\sim$ 4 Å). AFM images of CrO_2 films grown on these surfaces are shown in Figs. 2(c) and (d). The corresponding images at a lower magnification are shown in Figs. 2(e) and (f), respectively. The type A film, which is essentially strain-free, displays very square-like grains (Fig. 2(c) and 2(e)). On the other hand, the strained type B film exhibits more rectangular shaped grains, with the long direction being along [001]. This suggests that the lateral growth rates in the two directions are different for the growth of the strained films, with [001] being the faster growth direction. This is consistent with the results obtained on the selective growth of CrO_2 on patterned substrates, where a strong angular dependence of the lateral growth rate is observed [7]. The different shape grains of the type A and B samples suggest that the strain in the latter might influence the relative growth rates in the two directions. The smaller lattice mismatch in the [010] direction quite likely induces lateral growth in the normal direction, i.e. along [001]. For a very crude estimate, we can assume that the lateral growth rate is proportional to the strain anisotropy energy in that direction, an aspect ratio of 2.7 is thus expected. We observe an aspect ratio of about 3 for all the type B films, independent of thickness. This indicates that the strain in the two directions is relaxed at roughly the same rate with increasing thickness, which can also be inferred from Fig. 1.

On the contrary, in films that are essentially strain-free, no special growth direction is preferred, and one would expect a near unity grain aspect ratio, as is evidenced in the type A films.

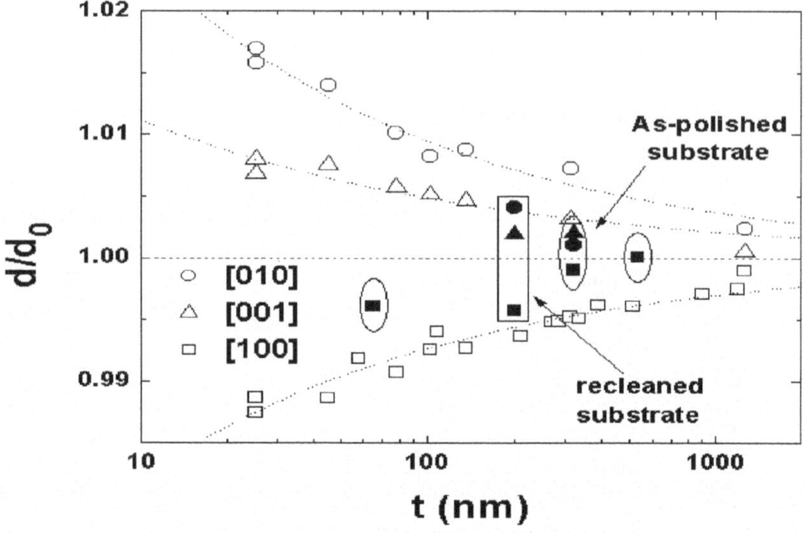

Fig. 1. Lattice parameters for epitaxial type B (open symbols) CrO_2 films as a function of thickness. Dashed lines are power law fittings, with the exponent being approximately 0.38. The lattice parameters for type A films of a few select thicknesses are shown as closed symbols

3.2 Magnetic Properties

Next, we focus our attention on the magnetic properties of the type A and B films. As previously noted, the type B films that grow on atomically ordered surfaces are much more heavily strained than the type A films. As we shall see this is directly reflected in their magnetic switching properties. For the strained films we describe the free energy of the system as:

$$E = K_0 + K_1 \sin^2\vartheta + K_2 \sin^4\vartheta + \left(K_{\sigma c}\sin^2\vartheta + K_{\sigma b}\cos^2\vartheta\right) = \\ = const + K_{1eff}\sin^2\vartheta + K_2\sin^4\vartheta \tag{1}$$

where K_1 and K_2 are the magnetocrystalline anisotropy energy constants; K_{sb} and K_{sc} are the strain anisotropy energy constants associated with the b- and c-axis directions, respectively; ϑ is the angle between the magnetization and the c-axis; K_{1eff} is the effective anisotropy energy constant K_{1eff} =

$K_1 + (K_{\sigma c} - K_{\sigma b}) = K_1 + \frac{3}{2}\lambda Y\,(\varepsilon_c - \varepsilon_b)$; λ is the magnetostriction coefficient; Y is the Young's modulus; and e is the strain [8]. Since the strain is larger in the b-axis direction than in the c-axis direction due to the larger lattice mismatch, the second term in the expression for K_{1eff} is negative. For very heavily strained films the effective anisotropy energy constant K_{1eff} can thus be negative, resulting in an easy axis reorientation towards the b-axis as has been experimentally observed [9].

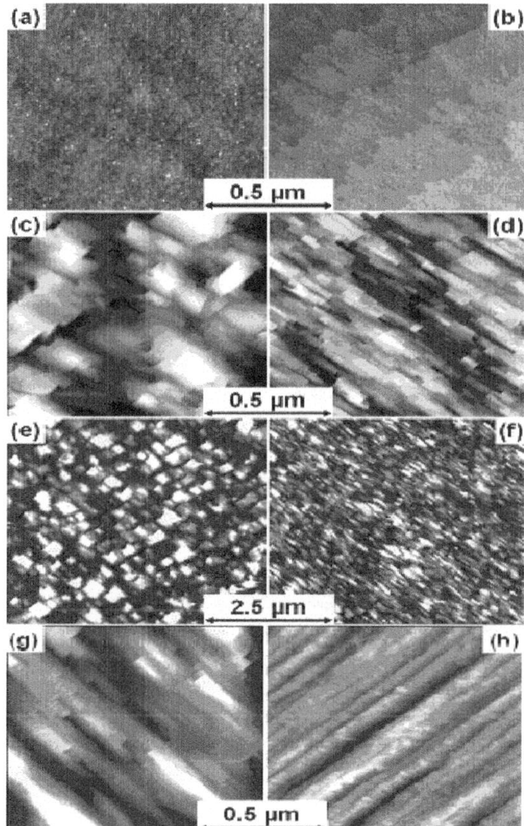

Fig. 2. Surface morphology as measured using AFM of the starting (100)-TiO_2 substrate surfaces and grown CrO_2 films. TiO_2 substrate (a) before and (b) after HF cleaning; (c) and (e) 65 nm type A CrO_2 film at different magnifications; (d) and (f) 37 nm type B film at different magnifications; (g) 85 nm type B film CrO_2 film; and (h) surface after chemical etching of film (g). The [001] and [010] directions of the substrate are marked in (a) and is the same in the other figures.

As seen in Fig. 1, the degree of strain in the type B epitaxial CrO_2 films is dependent on the thickness. For sufficiently thin films, the strain anisotropy can be larger than the crystalline anisotropy, and the easy axis tends to align along the b-axis. For thicker films, the crystalline anisotropy dominates and the easy axis is along the c direction as seen in Fig. 3. The critical thickness for the easy axis rotation has been determined to be at about 50 nm at room temperature. It should be noted that the values of both the crystalline and the strain anisotropy terms are temperature dependent. The former increase monotonically, almost doubling in value as the temperature is decreased from room temperature to liquid helium [10]. The strain anisotropy term is also temperature dependent through the changes of the magnetostriction coefficient, the Young's modulus and also the differences in the thermal expansion between the film and substrate. We find that an easy axis reorientation can also be achieved through changing the temperature as shown in Fig. 4, with the b-axis being preferred at low temperatures. Although measurements of the temperature dependence of the strain anisotropy for CrO_2 have not yet been reported, our results suggest that the temperature dependence of the product $\lambda Y \varepsilon$ is likely to be much stronger than that for K_1.

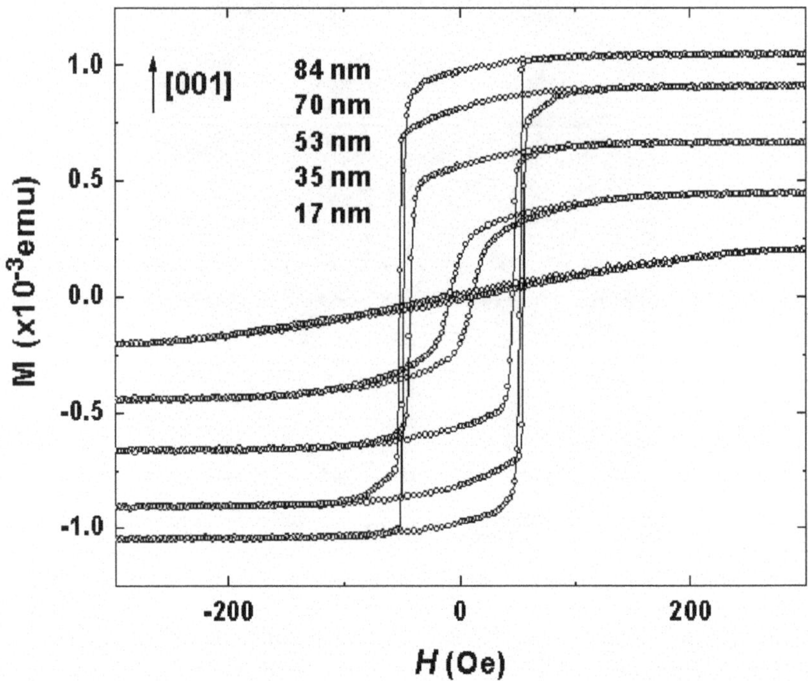

Fig. 3. Hysteresis loops displaying the evolution of the easy axis into the hard axis direction for type B (100)-oriented CrO_2 films with decreasing film thickness. The measurements for the all the films are at room temperature along the c-axis direction.

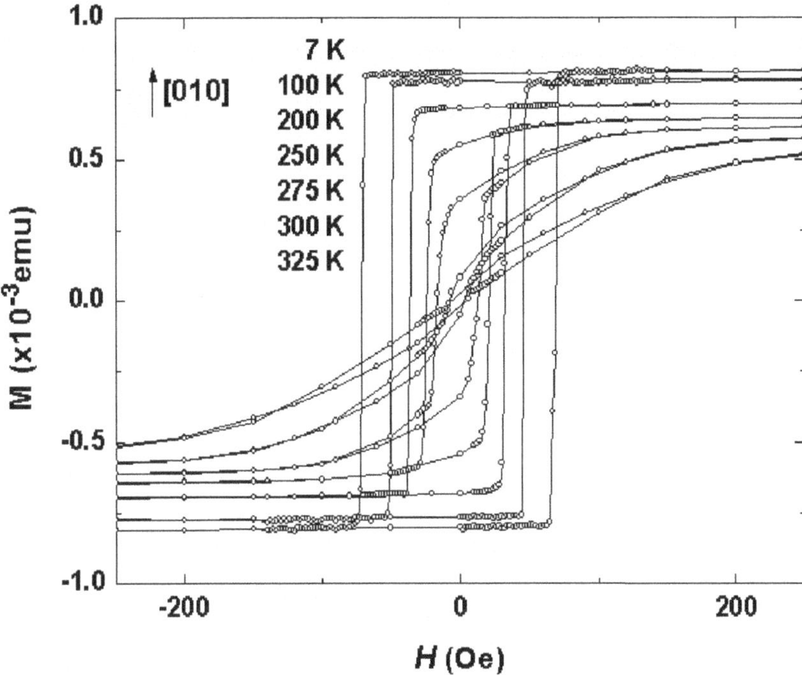

Fig. 4. Hysteresis loops displaying the evolution of the magnetic easy axis into the hard axis direction with increasing temperature for a type B film. The measurements are along the b-axis direction for a 47 nm film.

In the case of type A films, the substrate surface is rough on an atomic scale and the nucleation and growth of the film occurs quite randomly. Correspondingly, the strain anisotropy in these films is significantly smaller than the crystalline anisotropy. Because of the presence of a large number of defects the RMS roughness of these films are typically about twice as large as those of the type B films. The disorder is also reflected in the structural quality, with the (200) rocking curve width being about five times larger than that for the type B films. Despite their inferior crystalline quality, these films exhibit magnetic switching behavior close to that observed in bulk single crystals of CrO_2. Fig. 5 shows the hysteresis loops of a type A and type B film, both of which are nominally 65 nm in thickness. The double switching phenomena [11], resulting from non-uniform distribution of strain, which normally appear in the type B films of intermediate thickness, is not observed in the type A films. Furthermore, because of the lack of any significant influence of strain in the latter, the magnetic anisotropy is close to being uniaxial. The magnetization is also much more uniform magnetization, resulting in a larger nucleation field. We have also performed ferromagnetic resonance (FMR) studies [12] on the identical set of type A and B films and the results

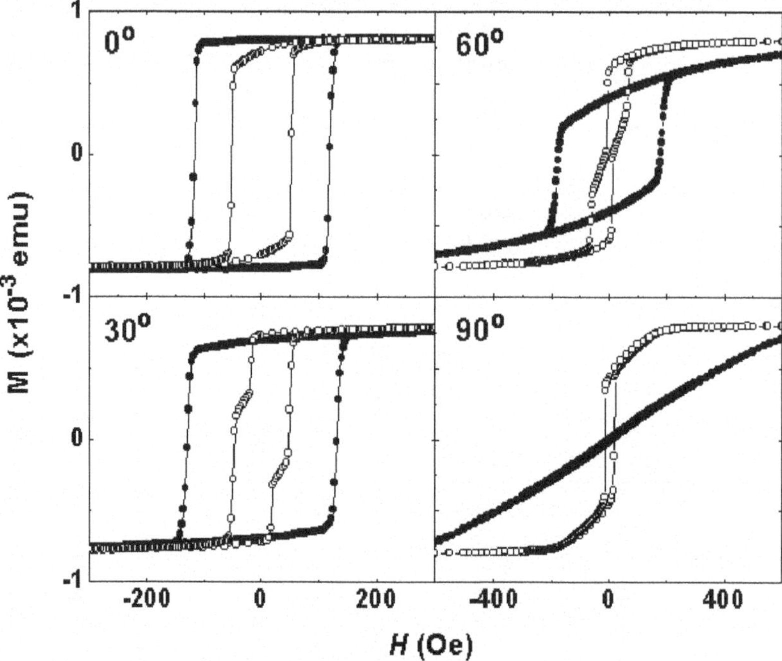

Fig. 5. Comparison of the hysteresis loops for a type A (open circles) and a type B (closed circles) CrO_2 film, both nominally 65 nm thick. The listed angles are with respect to the easy direction ([001] direction).

are consistent with the results reported here using conventional magnetic measurements.

As we show in Fig. 6, the Stoner-Wohlfarth model [13] provides a very good fit for the switching of type A films of different thickness up to the coercive fields. This suggests that the switching in these films occurs via coherent rotation that can be described as being close to single-domain like, and followed by domain wall motion above the nucleation fields. We have extracted the K_1 and K_2 values from the hard axis hysteresis loops and used them to generate the theoretical hysteresis curves for the other angles. The values used are, respectively, $K_1 = 13.7 \times 10^4$ erg/cm^3, $K_2 = 2.94 \times 10^4$ erg/cm^3 for the 65 nm film; and $K_1 = 22.1 \times 10^4$ erg/cm^3, $K_2 = 2.23 \times 10^4$ erg/cm^3 for the 535 nm film. The latter values are very close to the reported bulk CrO_2 anisotropy energy constants [10].

Figure 7 plots the angular dependence of the experimentally determined switching field, H_S, and the coercive field, H_C, for a 65 nm type B film. The Stoner-Wolhfarth model provides a reasonable fit for H_S only at high angles (i.e., close to the hard axis). On the other hand the Kondorskii relation [14,

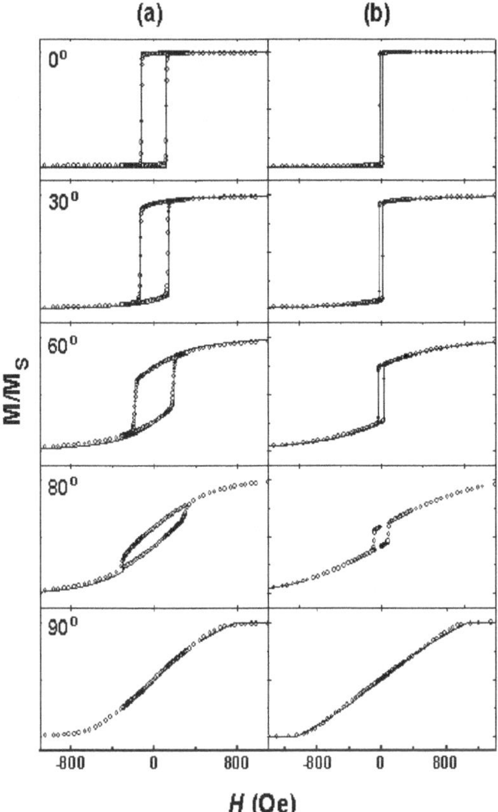

Fig. 6. Angular dependence of the hysteresis loops of type A CrO_2 films of: a) 65 nm and b) 535 nm thickness. The open circles are normalized experimental data points, while the solid lines are generated using the Stoner-Wohlfarth model except at the switching point. The K_1 and K_2 values for the fits at different angles are obtained from the hard-axis loop data.

15], $H_S(\varphi) = H_S(0)/ \mid cos\varphi \mid$, where φ is the angle between the applied field and c-axis, yields a more adequate fit at low angles. We find that by simply adding another parameter to this relationship, i.e.

$$H_S(\varphi) = H_S(0)\frac{b+1}{b+ \mid cos\varphi \mid} \tag{2}$$

where b is a fitting parameter, a satisfactory fit can be obtained over the whole range of angles. In addition to the HF surface treatment we have also investigated the effect of other TiO_2 substrate surface treatment procedures prior to the growth of CrO_2 films. This includes studying the influence of exposure to a low-energy ion beam. The films grown on ion beam irradiated

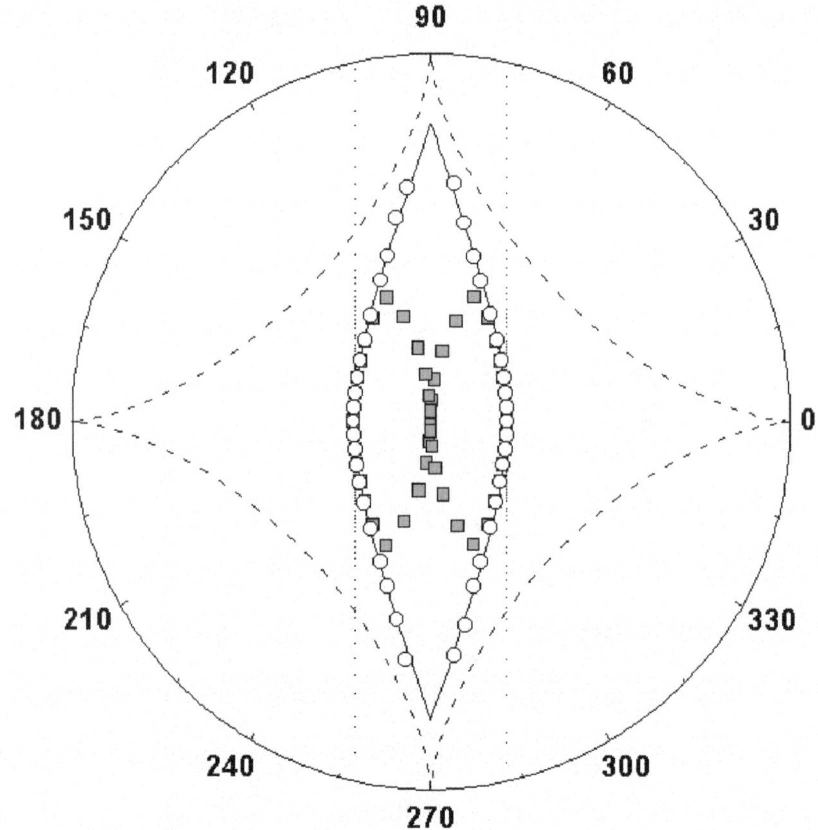

Fig. 7. Fitting of the angular dependence of the switching fields H_S (open circles) for a 65 nm type A CrO_2 film using three different models. Dashed line: Stoner-Wohlfarth model; Dotted line: Kondorskii model; Solid line: modified Kondorskii model, $H_S(\varphi) = H_S(0)\frac{b+1}{b+|cos\varphi|}$. The squares represent the experimentally determined coercive field values, H_C.

substrates are also strained, although not to the same extent as those on HF-treated substrates. A systematic study as a function of ion beam energy and exposure time is needed in order to better understand the microscopic influence of this surface treatment procedure.

The CrO_2 films grown on TiO_2 substrates can be readily etched off by chemical treatment using a standard chromium photomask etchant solution (e.g., from Cyantek Corp.). The cleaned substrates can then be reused (after HF pre-treatment) for subsequent growth of CrO_2 films. We find that repeatedly re-cleaned TiO_2 substrates also lead to the growth of strained CrO_2 films, but progressively less so with increasing usage as compared to virgin HF-treated substrates. This is not surprising considering that the sur-

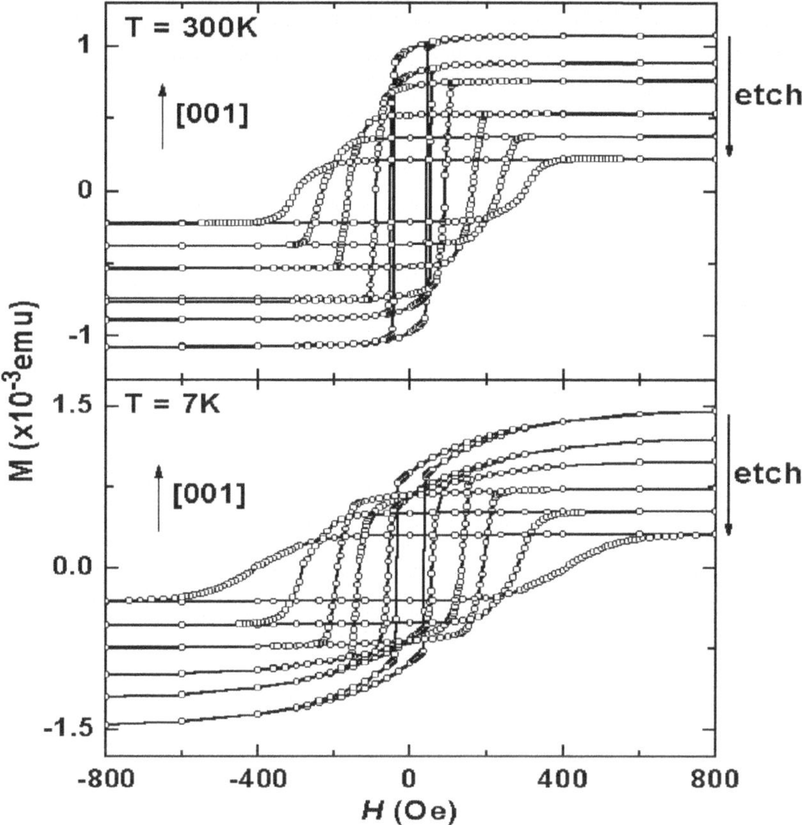

Fig. 8. Hysteresis loops at 300 K and 7 K along the c-axis direction as a function of increasing chemical etching of a CrO_2 film, with starting thickness of about 85 nm.

face becomes increasingly rougher with each deposition and surface cleaning cycle. An alternative method of modifying the CrO_2 magnetic properties is by chemical etching (using a dilute chromium photomask etchant solution) of as-deposited films. Figure 8 shows the hysteresis loops measured at 300 K and 7 K as a function of different etching time. The starting film is around 85 nm in thickness, and partial easy axis switching can be observed from its initial hysteresis loops. Chemically etching off the film gradually makes the hysteresis loops more and more square like in the sense that the magnetic remnance M_R/M_S becomes closer to unity with increasing etching. We find that after chemical etching, films as thin as 20 nm still retain their easy axis along the c-axis direction, even at low temperatures. In contrast, we have found that an as-grown 20 nm CrO_2 film will have its easy axis completely aligned along the b-axis, both at room temperature and at low temperatures [11]. Another important observation is that the coercive field increases with etching time and becomes significantly large for very thin layers. This behavior is not observed in the as-grown thin films of different thicknesses. We have noted previously that the weakly strained CrO_2 films exhibit a larger coercivity and anisotropy field than those that are more heavily strained. Thus, one possible explanation for the increase in the coercive field with etching is because of strain relaxation. Additionally, wet etching can modify the surface contribution to the total magnetization. As seen from the AFM image of an etched film (Fig. 2(h)), the etching process is very anisotropic. While the grains tend to preferentially align along the c-direction in the as-grown film (Fig. 2(g)), the film after etching is quite rough exhibiting periodic peaks and troughs that extend along the b-direction (Fig. 2(h)). This indicates the anisotropic nature of the etch process, with the etch rate being much faster in the a- and b-directions than in the c-direction.

Acknowledgment

We thank B. Z. Rameev, L. R. Tagirov, B. Aktas, and W. D. Doyle for valuable suggestions. This work was supported by the National Science Foundation Grant Nos. DMR-0306711 and DMR-0080031.

References

1. K. Schwarz, J. Phys. F. **16**, L211 (1986).
2. S. P.Lewis, P. B. Allen, T. Sasaki, Phys. Rev. B **55**, 10253 (1996).
3. M. A. Korotin, V. I. Anisimov, D. I. Khomskii, and G. A. Sawatzky, Phys. Rev. Lett. **80**, 4305 (1998).
4. Y. Ji, G. J. Strijkers, F. Y. Yang, C. L. Chien, J. M. Byers, A. Anguelouch, G. Xiao, and A. Gupta, Phys. Rev. Lett. **86**, 5585 (2001).
5. A. Anguelouch, A. Gupta, G. Xiao, D. W. Abraham, Y. Ji, S. Ingvarsson, and C.L. Chien, Phys. Rev. B **64**, 180408(R) (2001).
6. A. Gupta, X. W. Li and G. Xiao, J. Appl. Phys. **87**, 6073 (2000).
7. A. Gupta, X. W. Li, S. Guha, and G. Xiao, Appl. Phys. Lett. **75**, 2996 (1999).
8. B. D. Cullity, Introduction to Magnetic Materials (Addison-Wesley, London, 1972).
9. X. W. Li, A. Gupta, and G. Xiao, Appl. Phys. Lett. **75**, 713 (1999).
10. D. S. Rodbell, J. Phys. Soc. Japan **21**, 1224 (1966).
11. X. Miao, G. Xiao, and A. Gupta, Phys. Rev. B, in print.
12. B. Z. Rameev, A. Gupta, G. X. Miao, G. Xiao, F. Yildiz, L. R. Tagirov, and B. Aktas, Phys. Stat. Sol. (A) **201**, 2250 (2004).
13. E. C. Stoner, and E. P. Wohlfarth, Phil. Trans. Roy. Soc. A **240**, 599 (1948).
14. E. I. Kondorskii, J. Phys. SSSR **2**, 161 (1940).
15. W. D. Doyle, J. E. Rudisill, and S. Shtrikman, J. Phys. Soc. Jpn. **17**, 567 (1962).

Antiferromagnetic Interlayer Exchange Coupling Across Epitaxial Si Spacers

D.E. Bürgler, R.R. Gareev, L.L. Pohlmann, H. Braak, M. Buchmeier, M. Luysberg, R. Schreiber, and P.A. Grünberg

Institut für Festkörperforschung, Forschungszentrum Jülich GmbH,
D-52425 Jülich, Germany
D.Buergler@fz-juelich.de

Summary. We report on sizable antiferromagnetic interlayer exchange coupling (AFC) of Fe(001) layers across epitaxial Si spacers, for which epitaxial growth of a pseudomorphic phase stabilized by the interface is confirmed by low-energy electron diffraction and high-resolution transmission electron microscopy. The coupling strength decays with spacer thickness on a length scale of a few Å and shows a negative temperature coefficient. Transport measurements of lithographically structured junctions in current-perpendicular-to-plane geometry show the validity of the three "Rowell criteria" for tunneling: (i) exponential increase of resistance R with thickness of the barrier, (ii) parabolic dI/dV–V curves, and (iii) slight decrease of R with increasing temperature. Therefore, AFC is mediated by non-conductive spacers, which in transport experiments act as tunneling barriers with a barrier height of several tenths of an eV. We discuss our data – in particular the strength, thickness and temperature dependence – in the context of two previously proposed models for AFC across non-conducting spacers. We find that neither the molecular-orbital model for heat-induced effective exchange coupling nor the quantum interference model extended to insulator spacers by introducing complex Fermi surfaces can account for the strong AFC across epitaxial Si spacers and its negative temperature coefficient. The recently proposed defect-assisted interlayer exchange coupling model, however, yields qualitative agreement with the enhanced AFC and the temperature dependence.

1 Introduction

Magnetic interlayer exchange coupling across metallic spacer layers was discovered in 1986 by Grünberg et al. [1] and has been extensively investigated. It is well established that the coupling displays a damped oscillation between the ferro- and antiferromagnetic (AF) state as a function of the interlayer thickness [2]. Typical coupling strengths are of the order of 1 mJ/m^2. Theoretically, it was shown that the coupling across metals is due to the formation of standing electron waves in the interlayer, which result from spin-dependent electron interface reflectivity. When applying the same theoretical framework to insulating or semiconducting interlayers, however, Bloch states in the spacer have to be replaced by evanescent states, which exponentially decay with distance from the interfaces to the metallic, magnetic layers [3].

Accordingly, the coupling strength is also expected to exponentially decay when the thickness of a non-conducting interlayer increases. Furthermore, this model predicts an increase of the coupling strength with temperature for insulating spacers [3, 4] in clear contrast to metallic interlayers, where the model predicts and experiments confirm a decreasing coupling strength with temperature. The experimental data basis concerning coupling across non-metallic spacers is rather thin. For amorphous insulators like a-SiO_2 and a-Al_2O_3, which are widely employed e.g. in tunneling magneto-resistance devices, interlayer exchange coupling is not observed experimentally. However, there is a recent report of AF coupling with a strength of about 0.26 mJ/m^2 in epitaxial Fe/MgO/Fe(100) structures for very thin (< 7 Å) MgO thicknesses [5]. This report and our observation of even stronger antiferromagnetic interlayer exchange coupling (AFC) across nominally pure Si [6] focus particular interest on this new class of highly resistive structures exhibiting non-oscillatory AFC.

Previously, we have found that insulating-type, highly resistive Si spacers can be prepared by a certain deposition procedure [6, 7, 8]. Corresponding Fe/Si/Fe structures reveal very strong AFC with a total coupling strength in excess of 5 mJ/m^2 [6], which could be further increased to 8 mJ/m^2 by inserting thin epitaxial and metallic FeSi boundary layers at the spacer interfaces [8]. The just mentioned coupling strengths are among the strongest reported in literature including metallic spacers [2] and exceed the values obtained for metallic $Fe_{0.5}Si_{0.5}$ spacers grown by co-evaporation by one order of magnitude [7]. The thickness dependence of the coupling is oscillatory for metallic $Fe_{0.5}Si_{0.5}$ spacers, but exponentially decaying for Si-rich, highly resistive spacers. For combined semiconducting/metallic epitaxial spacers (i.e. nominally pure Si/$Fe_{0.5}Si_{0.5}$), the main impact to AFC originates from the semiconducting part of the spacer [9]. Finally, we also reported sizable AFC across epitaxial, Ge-containing spacers, when direct contact between Ge and Fe is prevented, e.g. by inserting thin Si boundary layers or by piling up thin layers of Ge and Si to form Si-Ge-multilayer spacers [10]. The latter results indicate that relatively strong AFC might be a common feature of well-ordered, epitaxial semiconducting spacer layers.

In order to clarify the coupling mechanism and to perform meaningful ab-initio calculations, detailed knowledge about the spacer layer in terms of structure as well as electronic properties is needed. High-resolution transmission electron microscopy (TEM) images as well as low-energy electron diffraction (LEED) are employed to study the crystalline structure of the Si interlayers, and we perform transport measurements with the current flowing perpendicular to the samples plane (CPP) to obtain additional and clear information whether Si-rich spacers are metallic or insulating. A further question is whether the transport in highly resistive spacers is due to elastic tunneling, or whether it arises from additional channels of conductivity across submicron-sized pinholes, as it was pointed out in [11, 12]. Pinholes also provide contacts between the FM layers where direct exchange interaction could strongly influence the coupling behavior. In fact, this extrinsic pinhole-induced coupling could obscure the intrinsic coupling mechanism. In order to address these questions we examine for epitaxial Fe/Si/Fe structures the validity of the necessary and sufficient Rowell criteria for direct elastic tunneling [12], i.e. (i) strong and exponential increase of the resistance R with spacer thickness t, (ii) parabolic dependence of conductivity versus bias voltage, and – most decisive – (iii) small and negative temperature coefficient of the zero-bias resistance [13]. Additionally, we measure the temperature dependence of the coupling as a further characteristic that can be compared to theoretical predictions for AFC across non-metallic spacers as presented in the concluding discussion.

2 Experimental Procedures

2.1 Sample Preparation

We grow our Fe/Si/Fe(001) structures in a molecular-beam epitaxy system using a 150 nm-thick Ag(001) buffer system on GaAs(001) [6, 7]. The layers forming the spacers are deposited at low deposition rates (< 0.1 Å/s) and at room temperature (RT). In some cases the spacers are grown in the shape of wedges to facilitate the study of thickness dependences. The nominal thickness of the wedges ranges from 8 to 20 Å over a lateral distance of typically 10 mm, and the Fe layer thicknesses lie in the range between 50 and 100 Å.

2.2 Magnetic and Structural Characterization

Magnetic properties are measured by magneto-optical Kerr effect (MOKE) in Voigt geometry, by magnetometry using a superconducting quantum interference device (SQUID), and Brillouin light scattering from spin waves (BLS). Bilinear (J_1) and biquadratic (J_2) coupling constants are determined by fitting the field dependence of MOKE, SQUID, and BLS data using the standard areal energy density expression

$$E_{ex} = -J_1 \cos(\vartheta) - J_2 \cos^2(\vartheta) \qquad (1)$$

to phenomenologically describe interlayer exchange coupling, where ϑ is the angle between the two Fe film magnetizations. The external magnetic field for all three techniques is applied along an easy-axis of Fe(001) in the plane of the sample. Further details concerning the preparation of the structures, their characterization, and the fitting procedures are described in [6, 7, 14].

The in-plane crystalline structure of all layers is characterized by means of in-situ LEED measurements. A TEM with aberration correction [15] is employed to obtain high-resolution images of Si spacer layers.

2.3 Lithography and Transport Measurements

The CPP transport measurements are performed after patterning 10×10 mm^2-sized, wedge-type samples using photolithography, ion-beam etching, and the lift-off technique. The layout of the patterned sample is shown in Fig. 1. In this way we obtain CPP junctions with different Si spacer thicknesses t and variable junction areas A, which all are deposited under the same growth conditions. We use crossed contacts, where a 300 nm-thick Cu layer forms the upper electrode. The patterned 150 nm-thick silver buffer layer serves as a bottom electrode. The sheet resistances of both electrodes are about 0.1 Ω and thus significantly smaller than the resistance of the tunneling junctions in CPP geometry (5–300 Ω), such that current distribution effects are diminished [16]. Insulation of the electrodes is achieved by deposition of a 250 nm-thick Si-oxide layer. Finally, we define junctions of rectangular shape ranging in area A from 22 to more than 200 μm^2. A photograph of a typical junction is shown in the inset of Fig. 1. After patterning, voltage and current leads suitable for four-point transport measurements are connected by ultrasonic bonding to measure the I–V characteristics of the junctions.

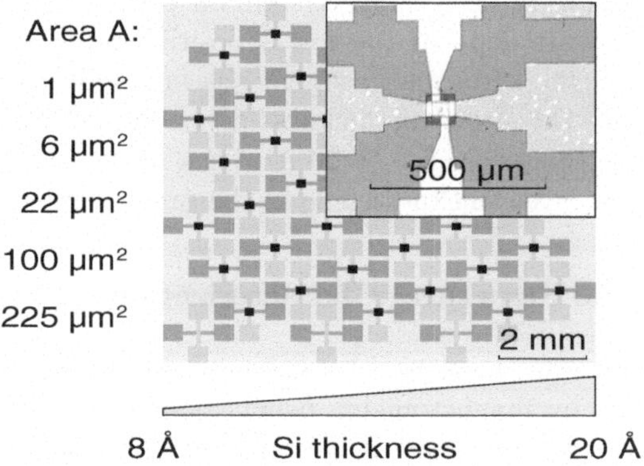

Fig. 1. Layout of the sample for CPP transport measurements. The Si spacer thickness t varies along the horizontal axis and the junction size A along the vertical axis. The inset shows a photograph of a patterned junction with the contact leads (white and light grey).

Examples of LEED patterns of a 5 nm-thick bottom Fe(001) layer and a 5 Å-thick Si interlayer grown on the top of the bottom Fe layer are shown in Fig. 2. Both pattern are taken at an electron energy of 55 eV and reveal the same surface reciprocal lattice in terms of symmetry, relative orientation, and lattice constants. The superimposed dashed squares connect the (01) spots and yield an in-plane lattice constant of 2.9 Å, the bulk value of bcc-Fe. Therefore, the in-plane structure of thin Si layers is the same as for the

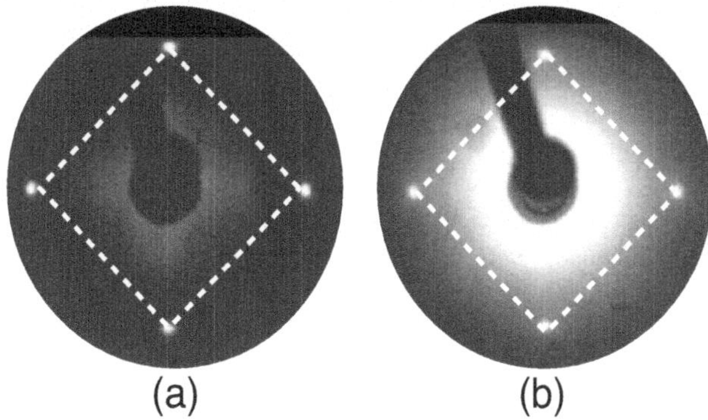

Fig. 2. LEED patterns of a 5 nm-thick Fe(100) bottom layer (a) and a 5 Å-thick Si spacer (b) grown at RT on the Fe layer shown in (a). Dashed squares mark the in-plane reciprocal lattice of bulk bcc-Fe(001) corresponding to an in-plane lattice constant of 2.9 Å.

Fe(001) surface. The LEED pattern of the top Fe(001) layer (not shown), i.e. the 5 to 10 nm-thick Fe layer grown on top of the Si spacer of Fig. 2(b), is very similar to the one of the bottom Fe layer shown in Fig. 2(a) and confirms the epitaxial growth throughout the whole stack.

Epitaxy is further confirmed by the TEM picture in Fig. 3(a), where it is indeed difficult to distinguish the Fe layers and the Si interlayer, because the atomic lattices match almost perfectly. A Fourier transform analysis of the vertical lattice distortion $\Delta g/g$ along the arrow in Fig. 3(a) reveals a dif-

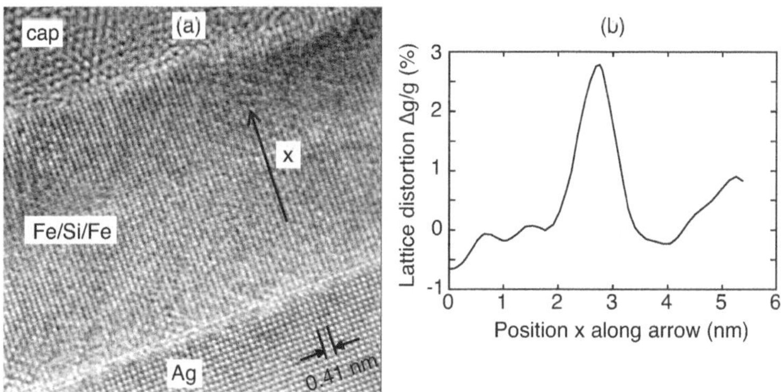

Fig. 3. (a) TEM image of a Fe/Si/Fe trilayer grown on Ag(001) (bottom right) and capped with a ZnS protection layer (top left). (b) Vertical lattice distortion $\Delta g/g$ along the arrow in (a), where g is the vertical separation of the atomic planes in the Fe layers.

ference of about 3% between the Fe layers and the Si interlayer [Fig. 3(b)]. The fact that the Si interlayer is vertically expanded with respect to the Fe lattice directly excludes that the Si grows in a tetragonally distorted bulk structure, because the in-plane expansion of the diamond lattice due to the slightly larger Fe(001) lattice would result in a vertical contraction of the order of 9% instead of an expansion by 3%. Therefore, the interlayer adopts a metastable, epitaxially stabilized structure, for which intermixing with Fe cannot be excluded solely based on the TEM data.

2.4 Thickness Dependence of AFC

The thickness dependence of the bilinear coupling strength J_1 of a Fe(50 Å)/Si(8–20 Å)/Fe(50 Å) trilayer is shown in Fig. 4. $|J_1|$ decays exponentially with t with a decay length of about 3 Å. For $t \approx 20$ Å the coupling strength decreases to $|J_1| \approx 0.1$ mJ/m^2. The zero-field antiparallel alignment is observed in the whole range of temperatures and for all spacer thicknesses. A typical experimental MOKE loop for a Si thickness of 17.3 Å (black) is shown in the inset of Fig. 4 together with the fit (grey circles) that yields antiparallel alignment at zero field (see arrows) due to a bilinear coupling strength of $J_1 = -0.27$ mJ/m^2.

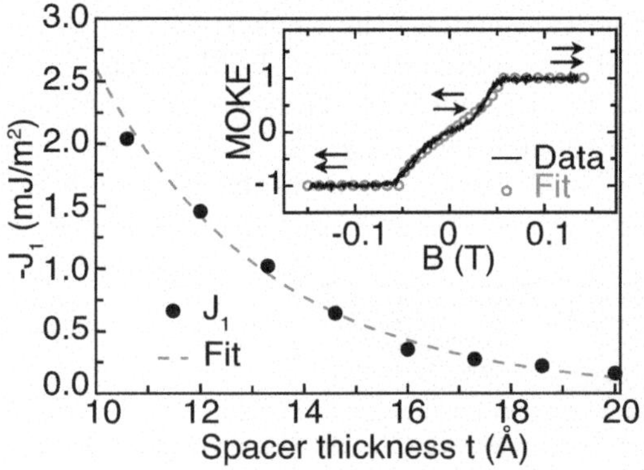

Fig. 4. Bilinear coupling constant J_1 of a Fe(50 Å)/Si(t)/Fe(50 Å) structure versus spacer thickness t measured at RT. The fitted curve yields a decay length of 3.3 Å. Inset: Experimental and fitted longitudinal MOKE hysteresis curves for $t = 17.3$ Å clearly show antiparallel alignment (arrows) due to AFC and yield $J_1 = -0.27$ mJ/m^2.

2.5 Temperature Dependence of AFC

Figure 5 shows the temperature dependence of the coupling across a Si spacer of 10 Å thickness yielding moderate coupling strength at RT. The Fe layer thickness of 100 Å is larger than our standard value to facilitate the analysis of the SQUID data. Magnetization loops measured by SQUID are fitted using the scheme described in [14], which takes into account the possibility of a twisted magnetization state due to the strong AFC and, thus, allows for an unequivocal separation and precise determination of J_1 and J_2. The independently, but for the same sample measured temperature dependence of the magnetization is also taken into account. The saturation magnetization drops from 10 to 300 K by about 20%. Both coupling parameters almost linearly decrease with increasing temperature. J_1 decreases from 10 to 300 K by almost 50% and J_2 by about 70%. This temperature dependence is of the same order of magnitude than what we have found previously [7] for metallic $Fe_{0.5}Si_{0.5}$ spacer layers. There, the total coupling at the second oscillation maximum, which is dominated by bilinear coupling, decreases from 80 to 300 K by about 45% (again taking into account a drop of the saturation magnetization by about 20%) and levels off below 80 K. The grey solid line in Fig. 5 is the prediction of the quantum interference model [3] and will be discussed in Sect. 3.

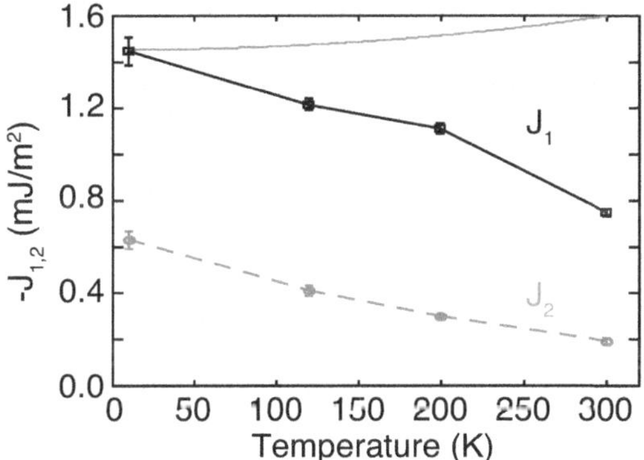

Fig. 5. Bilinear and biquadratic coupling constants J_1 and J_2 of a Fe(100 Å)/Si(10 Å)/Fe(100 Å) structure versus temperature T. The values are derived from fitting SQUID magnetization loops. The solid grey line is the temperature dependence of J_1 predicted by the quantum interference model for insulating spacers via equation (2).

2.6 Transport Measurements

First Rowell Criterion

In Fig. 6 we show the resistance times area product RA versus t on a semilogarithmic scale. The value of RA increases at RT strongly with t by more than 4 orders of magnitude, while t only approximately doubles. The characteristic length t_0 of the order of 1 Å (dashed line in Fig. 6) is significantly shorter compared to previously reported values for structures with amorphous Si spacers [17]. Note, that the coupling strength in Fig. 4 decays with a decay length of the same order of magnitude as the tunneling conductivity.

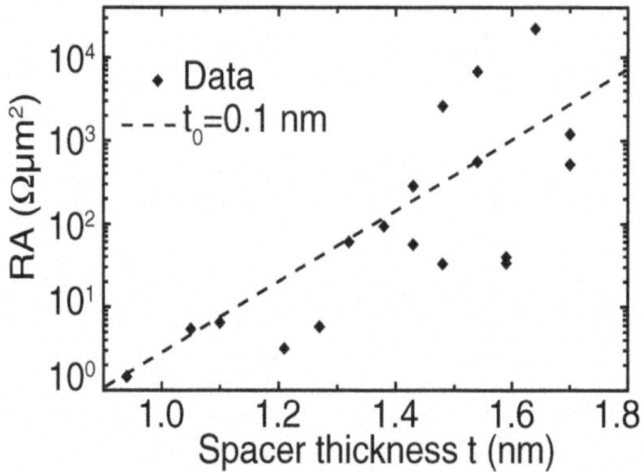

Fig. 6. First Rowell criterion: Dependence of the resistance times area product RA on the nominal spacer thickness t obtained from Fe(50 Å)/Si(t)/Fe(50 Å) junctions with areas A between 22 and 225 μm^2. The dashed line corresponds to a characteristic length $t_0 = 1$ Å.

Second Rowell Criterion

A representative I–V curve taken at RT and the corresponding dI/dV–V curve are presented in Fig. 7. They show the typical tunneling-type behavior. The dI/dV–V curve is parabolic with its minimum away from $V = 0$. These features are characteristic for tunnel junctions with asymmetric barriers and indicate different conditions at the diffused Fe/Si and Si/Fe interfaces [6, 8, 18]. There is no evidence for a conductivity anomaly near $V = 0$, as previously reported for ferromagnetic junctions with Al-oxide spacers and related to inelastic scattering assisted by magnons and impurities [19]. Similar I–V curves can occur when transport is due to another conductivity channel, namely submicron-sized pinholes, which can mimic elastic tunneling [11]. As we will show below based on an analysis of the temperature dependence of the resistance, this metallic-type channel gives here no significant contribution. We observe tunneling-type I–V curves only for $t > 15$ Å, where the voltage drop is sufficient to reveal the non-linear part of I–V characteristics. The barrier heights Φ derived from Brinkman fits [20] vary from 0.3 to 0.8 eV for different junctions, which all show a definite barrier asymmetry $\Delta\Phi$ in the

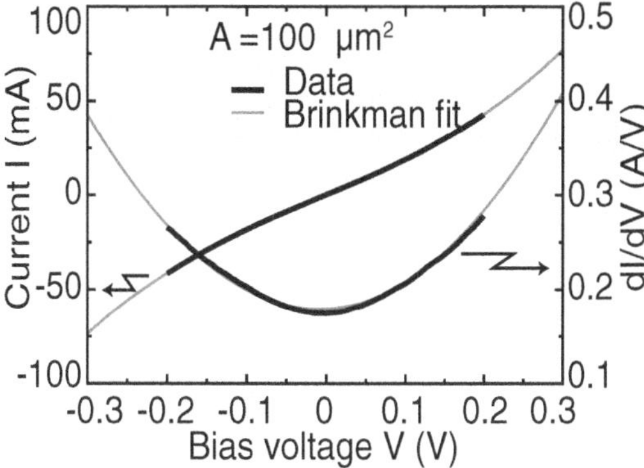

Fig. 7. Second Rowell criterion: Measured (black) and fitted (grey) I–V and dI/dV–V curves of a Fe/Si/Fe junction with $A = 100\ \mu m^2$ and $t = 15.4$ Å.

range from 0.1 to 0.3 eV. Explicit Brinkman fit results for a series of different junctions and a detailed discussion can be found in [13].

Third Rowell Criterion

A typical temperature dependence of the zero-bias resistance is presented in Fig. 8. The resistance slightly decreases with temperature and, thus, shows tunneling-type behavior. The total change of resistance from 4 K to RT does not exceed 5–7%. We relate the change of resistance to prevailing direct elastic tunneling, which yields only weak temperature dependence due to the

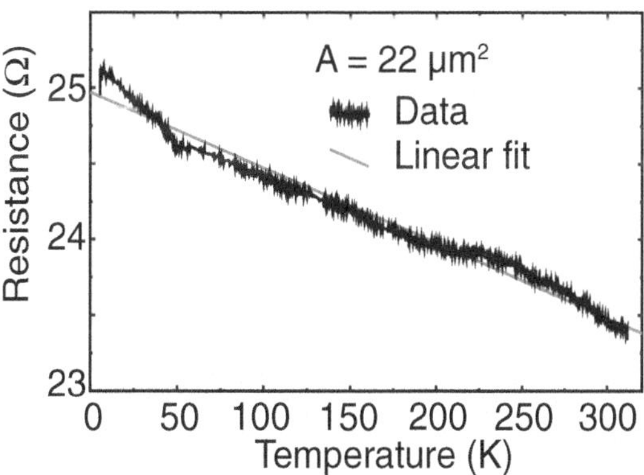

Fig. 8. Third Rowell criterion: Dependence of the resistance on temperature for a Fe/Si/Fe junction with $A = 22\ \mu m^2$ and $t = 17.0$ Å. The solid line is a linear fit that yields a temperature coefficient of $-5 \times 10^{-3} \Omega K^{-1}$.

broadening of Fermi distributions. The elastic but resonant tunneling channel is much weaker than the direct one and obeys a decay length, which is twice as large as the decay length of direct tunneling. However, resonant elastic tunneling cannot definitely be excluded for our junctions with t lying in the narrow interval between 14 and 17 Å. Different weights of the contributions from elastic direct and elastic resonant tunneling could lead to the scattering of the RA values in Fig. 6. Next, we consider inelastic tunneling based on thermo-activated hopping across impurity states in the barrier. For this channel a strong decrease of resistance with temperature is expected [17]. Thus, this channel is not dominant in our junctions.

3 Discussion

The transport measurements show that epitaxial, AF-coupled Fe/Si/Fe junctions fulfill all three necessary and sufficient Rowell criteria for direct elastic electron tunneling. A significant metallic contribution to the electron transport through pinholes can be excluded. The experimentally proven coexistence of strong AFC and electron transport via direct tunneling across nominally pure Si spacers proves that a non-conducting interlayer mediates the AF exchange coupling. The underlying, presumably so far disregarded coupling mechanism gives rise to a rather strong interaction, as the observed coupling strengths are among the largest ever reported – including metal spacers [2] – and clearly exceed the corresponding values for amorphous Si spacers by 3 orders of magnitude [21]. In the case of nominally pure Si spacers, the coupling also exceeds the values previously found for various $Fe/Fe_{1-x}Si_x/Fe$ structures [6, 7, 22, 23, 24].

Hunziker and Landolt [25] proposed a heat-induced coupling mechanism to explain the interlayer coupling across amorphous semiconductor spacers (a-Si, a-Ge, a-ZnSe), which is based on the interaction of localized, weakly bound states at the interfaces. These states are assumed to originate from impurities in the semiconductor material. They overlap in the spacer to form large molecular orbitals, for which the Pauli principle requires a different energy for the parallel and antiparallel spin configurations. This difference determines the coupling strength. A key feature of this mechanism is a strong positive temperature coefficient, which arises from the thermal population of these orbitals. For our epitaxial system, the transport measurements and the negative temperature coefficient in Fig. 5 negate heat-induced effects. Furthermore, a rather high density of impurities of the order of 10^{19} cm^{-3} must be assumed to obtain a 10^3 times stronger coupling than in [25]. Therefore, we dismiss this mechanism for epitaxial Fe/Si/Fe trilayers.

Another coupling mechanism for insulating spacers was derived by Bruno [3, 4] who extended the quantum interference model to insulating materials by introducing the concept of complex Fermi surfaces. Here, the coupling arises from spin-dependent interferences of electron waves – Bloch waves for metals and evanescent waves for insulators – in the spacer, which result from spin-dependent reflections at the interfaces. The model predicts for insulating (metallic) spacers a positive (negative) temperature coefficient, in both cases due to the thermal smearing of the Fermi surface. For metals the fuzziness of the Fermi surface affects the interference condition, and for insulators states above the Fermi level experience a lower tunneling barrier and, thus, have a higher transmission probability. The temperature dependence of J_1 for an insulating spacer is given by [3, 4]

$$J_1(T) = J_1(0)\frac{T/T_0}{\sin(T/T_0)}, \tag{2}$$

where T_0 is of the order of 200–600 K for barrier heights of 0.1–0.9 eV [3, 4]. Therefore, the coupling is almost constant below 300 K as demonstrated by the solid grey line in Fig. 5 which is calculated for $T_0 = 400$ K corresponding to a barrier height of about 0.4 eV and normalized to the experimental J_1 value at 10 K. The experimental decrease of J_1 in Fig. 5 is in disagreement with the prediction of (2). Nevertheless, we consider the $T = 0$ limit, where Bruno's model reduces to Slonczewski's spin-current model [26], in order to compare the thickness dependences of experiment and model. Using a two-band approximation for the exchange-split ferromagnet, the coupling strength is given by [26]

$$J_1 = \frac{(U - E_{\mathrm{F}})}{8\pi^2 t^2}\frac{8k^3(k^2 - k_\uparrow k_\downarrow)(k_\uparrow - k_\downarrow)^2(k_\uparrow + k_\downarrow)}{(k^2 + k_\uparrow^2)^2(k^2 + k_\downarrow^2)^2}e^{-2kt}, \tag{3}$$

where $(U - E_{\mathrm{F}})$ is the barrier height, $k_{\uparrow(\downarrow)}$ the Fermi wave vectors for the spin up (down) bands of the ferromagnet, and

$$k^2 = \frac{2m_{\mathrm{eff}}(U - E_{\mathrm{F}})}{\hbar} \tag{4}$$

with m_{eff} the effective electron mass in the interlayer. This equation was employed by Faure-Vincent et al. [5] to successfully fit the strength and thickness dependence of the AF coupling in epitaxial Fe/MgO/Fe structures. If we apply the same procedure with the same parameters for Fe [27],

$$k_\uparrow = 1.09\ \mathrm{\AA}^{-1} \quad \text{and} \quad k_\uparrow = 0.43\ \mathrm{\AA}^{-1}, \tag{5}$$

to our data, then we get curves similar to the dashed line in Fig. 4, but the fitted values for the barrier height and the effective mass are physically not meaningful, e.g. several keV for $(U - E_{\mathrm{F}})$ and 10^{-5} rest masses for m_{eff}. The reason is the strong coupling which requires in (3) a large prefactor [i.e. a huge barrier height $(U - E_{\mathrm{F}})$]. On the other hand, the decay length in the exponent of (3) given by k must be of the order of a few Å and, thus, forces m_{eff} to be extremely small to compensate the huge $(U - E_{\mathrm{F}})$. In other words, the quantum interference or spin-current model, respectively, in the two-band approximation as the basis of (3) can for Fe/Si/Fe – in contrast to Fe/MgO/Fe in [5] – not at all account for the observed AF coupling.

Recently, Zhuravlev et al. [28] proposed a coupling model based on localized impurity or defect states within the tunneling barrier. In contrast to the model of Hunziker and Landolt [25], defects or impurities in the center of the barrier are considered because they contribute strongest. The resonant origin makes the defect-assisted coupling much stronger than that in the absence of defects. If the energy level of the defect states matches with the Fermi level of the ferromagnets, the coupling becomes antiferromagnetic and the coupling strength decreases with increasing temperature. All these features qualitatively agree with the experimental findings. Although we cannot preclude the presence of defect or impurity states in our nominally pure Si spacers, a defect concentration of several percent predicted by the model in order to reproduce the strong AFC in Fe/Si/Fe seems unrealistically high. On the other hand, the simple model describes the electronic structure of the ferromagnet by free-electron spin-split bands, the barrier by a rectangular potential, and the defect state by a delta-function. Therefore, it certainly has a limited quantitative prediction power; but the physics might be correct.

4 Conclusions

The very strong antiferromagnetic interlayer exchange coupling in epitaxial Fe/Si/Fe(001) trilayers is mediated by a non-conductive Si spacer layer that acts in CPP transport measurements as a tunneling barrier with a height of several tenths of an eV. The temperature dependence of the bilinear coupling constant determined taking into account the experimental temperature dependence of the saturation magnetization reveals a negative temperature coefficient. This behavior and the strength of the coupling are not compatible with the molecular-orbital model of heat-induced exchange coupling proposed for amorphous semiconductor spacer [25]. The quantum interference model [3, 4, 26] predicts the observed thickness dependence, but for our circumstances (i.e. barrier height) a rather weak positive temperature dependence, which is not compatible with the present data of a moderately coupled Fe(100 Å)/Si(10 Å)/Fe(100 Å) trilayer. Furthermore, the model in the simple two-band approximation for the ferromagnet and a "free-electron-like" tunneling behavior in the spacer completely fails to predict the observed coupling strength by at least one order of magnitude. Presently, the defect-assisted interlayer exchange coupling model by Zhuravlev et al. [28] yields the best qualitative description of the experiments, as it correctly predicts strong antiferromagnetic coupling with a negative temperature coefficient. However, the detailed and quantitative understanding of the mechanism for the strong antiferromagnetic coupling across epitaxial, highly resistive Si spacers still remains an open question.

References

1. P. Grünberg, R. Schreiber, Y. Pang, M. B. Brodsky, and H. Sowers, Phys. Rev. Lett. **57**, 2442 (1986).
2. D. E. Bürgler, P. Grünberg, S. O. Demokritov, and M. T. Johnson, in *Handbook of Magnetic Materials*, edited by K. H. J. Buschow (Elsevier, Amsterdam, 2001), Vol. 13, Chap. Interlayer Exchange Coupling in Layered Magnetic Structures, pp. 1–85.
3. P. Bruno, Phys. Rev. B **52**, 411 (1995).
4. P. Bruno, J. Appl. Phys. **76**, 6972 (1994).
5. J. Faure-Vincent, C. Tiusan, C. Bellouard, E. Popova, M. Hehn, F. Montaigne, and A. Schuhl, Phys. Rev. Lett. **89**, 107206 (2002).
6. R. R. Gareev, D. E. Bürgler, M. Buchmeier, R. Schreiber, and P. Grünberg, J. Magn. Magn. Mater. **240**, 237 (2002).
7. R. R. Gareev, D. E. Bürgler, M. Buchmeier, D. Olligs, R. Schreiber, and P. Grünberg, Phys. Rev. Lett. **87**, 157202 (2001).
8. R. R. Gareev, D. E. Bürgler, M. Buchmeier, R. Schreiber, and P. Grünberg, Appl. Phys. Lett. **81**, 1264 (2002).
9. R. R. Gareev, D. E. Bürgler, M. Buchmeier, R. Schreiber, and P. Grünberg, Trans. Magn. Soc. Japan **2**, 205 (2002).
10. R. R. Gareev, D. E. Bürgler, R. Schreiber, H. Braak, M. Buchmeier, and P. A. Grünberg, Appl. Phys. Lett. **83**, 1806 (2003).
11. D. A. Rabson, B. J. Jönsson-Åkerman, A. H. Romero, R. Escudero, C. Leighton, S. Kim, and I. K. Schuller, J. Appl. Phys. **89**, 2786 (2001).
12. J. J. Åkerman, R. Escudero, C. Leighton, S. Kim, D. A. Rabson, R. W. Dave, J. M. Slaughter, and I. K. Schuller, J. Magn. Magn. Mater. **240**, 86 (2002).
13. R. R. Gareev, L. L. Pohlmann, S. Stein, D. E. Bürgler, and P. A. Grünberg, J. Appl. Phys. **93**, 8038 (2003).
14. M. Buchmeier, B. K. Kuanr, R. R. Gareev, D.E. Bürgler, and P. Grünberg, Phys. Rev. B **67**, 184404 (2003).
15. C. L. Jia, M. Lentzen, and K. Urban, Science **299**, 870 (2003).
16. R. J. M. van de Veerdonk, J. Novak, R. Meservey, J. S. Moodera, and W. J. M. de Jonge, Appl. Phys. Lett. **71**, 2839 (1997).
17. Y. Xu, D. Ephron, and M. R. Beasley, Phys. Rev. B **52**, 2843 (1995).
18. R. Kläsges, C. Carbone, W. Eberhardt, C. Pampuch, O. Rader, T. Kachel, and W. Gudat, Phys. Rev. B **56**, 10801 (1997).
19. J. S. Moodera and G. Mathon, J. Magn. Magn. Mater. **200**, 248 (1999).
20. W. F. Brinkman, R. C. Dynes, and J. M. Rowell, J. Appl. Phys. **41**, 1915 (1970).
21. B. Briner and M. Landolt, Phys. Rev. Lett. **73**, 340 (1994).
22. Y. Endo, O. Kitakami, and Y. Shimada, J. Appl. Phys. **87**, 6836 (2000).
23. E. E. Fullerton, J. E. Mattson, S. R. Lee, C. H. Sowers, Y. Y. Huang, G. Felcher, S. D. Bader, and F. T. Parker, J. Magn. Magn. Mater. **117**, L301 (1992).
24. J. J. de Vries, J. Kohlhepp, F. J. A. den Broeder, R. Coehoorn, R. Jungblut, A. Reinders, and W. J. M. de Jonge, Phys. Rev. Lett. **78**, 3023 (1997).
25. M. Hunziker and M. Landolt, Phys. Rev. Lett. **84**, 4713 (2000).
26. J. C. Slonczewski, Phys. Rev. B **39**, 6995 (1989).
27. V. Moruzzi, *Calculated Electronic Properties of Transition Metal Alloys* (World Scientific, Singapore, 1994).
28. M. Y. Zhuravley, E. Y. Tsymbal, and A. V. Vedyayev, Phys. Rev. Lett. **94**, 026806 (2005).

Magnetic Tunneling Junctions – Materials, Geometry and Applications

G. Reiss[1], H. Koop[1], D. Meyners[1], A. Thomas[1], S. Kämmerer[1], J. Schmalhorst[1], M. Brzeska[1], X. Kou[1], H. Brückl[2], and A. Hütten[3]

[1] University of Bielefeld, Department of Physics, Nanodevice group, P.O. Box 100 131, 33501 Bielefeld, Germany
[2] MIT, Francis Bitter Magnet Lab., NW 14-2128, 170 Albany St., 02139 Cambridge, MA, USA
[3] ARCS research GmbH Nanosystems Technology, Tech Gate Vienna, Donau-City-Str. 1, 1220 Wien, Austria
[4] Research Center Karlsruhe GmbH, Institute for Nano-Technology, PO Box 3640, 76021 Karlsruhe, Germany

Summary. The discoveries of antiferromagnetic coupling in Fe/Cr multilayers by Grünberg, the Giant MagnetoResistance by Fert and Grünberg and a large tunnelling magnetoresistance at room temperature by Moodera have triggered enormous research on magnetic thin films and magnetoelectronic devices. Large opportunities are especially opened by the spin dependent tunnelling resistance, where a strong dependence of the tunnelling current on an external magnetic field can be found. Within a short time, the quality of these junctions increased dramatically. We will briefly address important basic properties of these junctions depending on the material stacking sequence of the underlying standard thin film system with special regard to complex interdiffusion properties. New materials with potentially 100% spin polarization will be discussed using the example of the full Heusler compound Co_2MnSi, where we obtain up to 100% TMR at low temperature. Next, we discuss scaling issues, i.e. the influence of the geometry of small tunnelling junctions especially on the magnetic switching behaviour down to junction sizes below 0.01 μm^2. The last part will give a short overview on field programmable logic circuits made from magnetic tunnelling cells, where we demonstrate the clocked operation of a programmed AND gate.

Introduction

In recent years the interest in magnetic tunnel junctions (MTJs) has increased due to their high potential as memory cells in magnetic random access memories (MRAMs) or read heads in hard disk drives [1, 2, 3, 4]. Nevertheless the magnetic switching behaviour of MTJs with lateral extensions below one micron is not yet understood in detail. Distorted switching curves (astroids) obtained from magnetoresistance curves were reported by, e.g., Klostermann et. al. [5]. Moreover, identically prepared tunnel junctions often show different junction specific switching behaviour [6, 7, 8]. On the one hand the physical origin of these variations is unknown up to now, on the other hand they limit the technical applicability of the MTJs. In this work we present investigations of sub-μ magnetic tunnel junctions. First we will discuss the influence of the stacking sequence on the magnetic and the related Tunnelling MagnetoResistance (TMR) properties. Using Auger electron spectroscopy, X-ray absorption and X-ray Magnetic Circular Dichroism, an improved understanding of the complex interdiffusion behaviour of the standard film stacks can be obtained and used for the interpretation of the development of the TMR with special regard to the annealing used for initiating the exchange bias with a natural antiferromagnet having Mn as one of the components [9].

A detailed study by atomic and magnetic force microscopy (MFM) [8], in combination with micro magnetic numerical simulations tries to give a deeper insight in the properties shown by individual MTJs. The lithographic steps in the fabrication process inevitably lead to imperfect boundaries of the MTJs with a roughness on the nanometer scale. The impact of these structural imperfections on the magnetic switching behaviour will be discussed. On MTJs smaller than 200 nm, the resolution of the MFM of around 30 nm and the small thickness of the ferromagnetic electrodes hinders a reliable imaging of the magnetic states during switching the soft electrode. On such small patterns, we therefore employed conducting force microscopy [6], (c-AFM), where we form a contact between the tip and the top electrode. With this technique, minor loops and complete astroids can be obtained on MTJs as small as around 50 nm.

The next section will address new materials integrated into MTJs. As an example for electrodes, we chose Co_2MnSi. As full Heusler alloy, this is one of the materials with possibly large spin polarization [10]. Our investigations, that this material can be used as ferromagnetic electrode; the values of the low temperature TMR reach about 100% indicating the potential of this alloy [11]. At room temperature, however, the TMR effect is still below the numbers obtained with conventional ferromagnets. Reasons for that will be discussed and possible improvements suggested.

With regard to the tunnelling barriers, MgO is a material with obviously larger potential than Al_2O_3 [12, 13, 14]. We will show first results on the integration of MgO barriers in tunnelling stacks with CoFeB electrodes and discuss physical properties of these novel MTJs.

The paper will close with the discussion of possible applications beyond MRAM's in the related logic circuits. In field programmable logic circuits made from magnetic tunnelling cells, we demonstrate the clocked operation of a programmed AND gate [15]. The possibility to produce logic gate arrays and memory on the common technology platform of magnetic tunnelling cells could be of major importance for the further development of magnetoelectronics.

1 The Development of Standard MTJ Cells

1.1 Stacking Sequence and Interdiffusion

First attempts to form reliable MTJ's tried to use one relatively hard- and one relatively softmagnetic ferromagnetic electrode. This, however, turned out to be not stable with respect to magnetically cycling the soft electrode, because the domain splitting of the soft electrode causes large stray fields which induce a deterioration of the hard magnetic material [16]. The same is true, if the hard electrode is additionally stabilized by an antiferromagnetically coupled trilayer as, e.g., Co / Cu / Co [17].

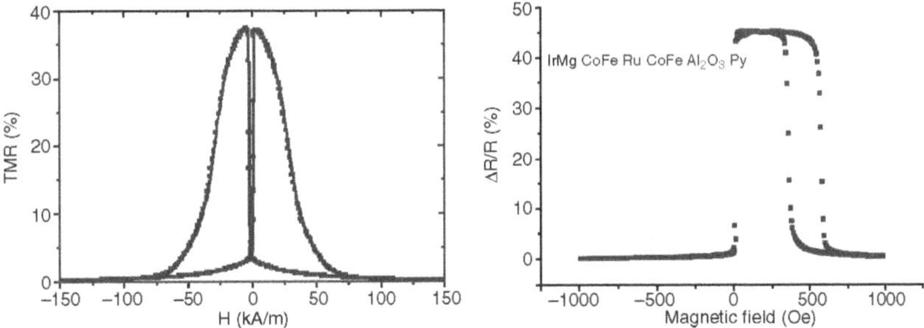

Fig. 1. Resistance as a function of an external magnetic field (major loop) for a tunnelling junction $Co/Cu/Co/Al_2O_3/Ni_{80}Fe_{20}$ from 1999 (left) and for a tunnelling junction IrMn/CoFe/ $Al_2O_3/Ni_{80}Fe_{20}$ from 2002 (right)

This typically leads to major loops as shown in Fig. 1 [18]. Moreover, the Cu turned out to diffuse to the barrier between Co and Al_2O_3 giving rise to a rapid decrease of the TMR while maintaining "good" tunnelling properties. Thus the nowadays standard stack design is a sequence of conductor- and seed layers followed by a natural antiferromagnet used for exchange biasing the following artificial antiferromagnet which mostly is a CoFe/Ru/CoFe trilayer in the second maximum of the antiferromagnetic coupling. This coupling further stabilizes the hard exchange biased electrode and gives the possibility to tailor its net magnetic moment. The tunnelling barrier is usually made by depositing an Al film with a thickness between 0.6 nm and 1.2 nm which then is oxidized by a mild treatment in an oxygen- or an Ar-O plasma. From intensive investigations it is known that the energy of the ions impinging on the film during this process should be kept well below 50 eV in order to avoid O implantation in the underlying ferromagnet. The next film is the soft magnetic electrode made of, e.g., CoFe, NiFe or a double layer of these alloys giving rise to a TMR of up to around 60%. Recently, values of 70% have been reported for an amorphous CoFeB electrode; the reason for this large number being not yet known causes some uncertainty concerning the maximum TMR reachable with conventional 3-d ferromagnets. This top soft ferromagnet is then covered again by a system of conductors, seeds, diffusion barriers and protecting layers.

The next step in the preparation routine is the initialization of the exchange bias which is usually done by annealing the film stack in a magnetic field at a typical temperature around 300°C for some minutes.

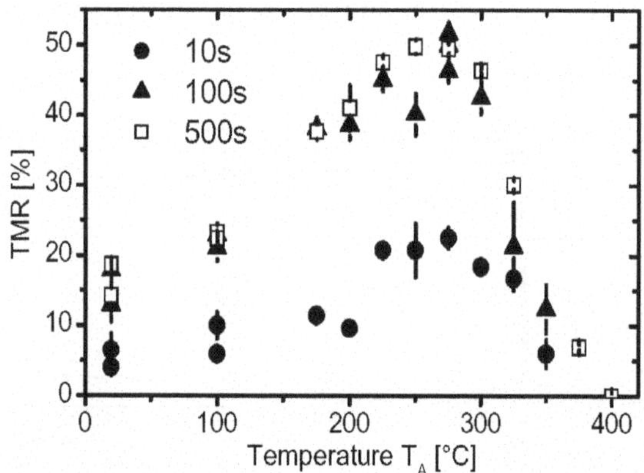

Fig. 2. Dependence of the TMR on the annealing temperature for a tunnelling film stack with IrMn and CoFe as hard and $Ni_{80}Fe_{20}$ as soft electrode for different oxidation times in a O_2 plasma

Figure 2 shows the dependence of the TMR on the annealing temperature for different oxidation times. Here, the oxidation was done by a remote ECR oxygen plasma, where 100s were found to be the optimum duration time. It is obvious, that the heavily underoxidized samples (10s in Fig. 2) have much lower TMR due to the remaining metallic Al. Surprisingly, the overoxidized sample (500s) does not show much lower TMR after annealing, although the other characteristics like symmetry of the I/V curve, breakdown voltage and bias voltage dependence of the TMR are considerably deteriorated compared with the samples oxidized for 100 sec. For all oxidation conditions the TMR increases up to a maximum around 275°C. For higher temperature the TMR decreases rapidly, with the decrease being weakest for $t_{Ox} = 500$ s (a similar result was observed for CoFe / AlO_x / Ni-Fe MTJs before [19]).

Here, the increase results partly from setting the exchange bias of the Co-Fe electrode by annealing in the magnetic field. In the as-prepared state the remanent magnetization of the Co-Fe layer is typically $M_R/M_S = 0.2$. This low remanent magnetization disables an antiparallel orientation of the electrodes resulting in a nearly 50% reduction of the TMR amplitude. For annealing temperature $T_A = 175°C$, the remanent Co-Fe magnetization is homogenous and oriented parallel to the field and a well defined antiparallel configuration of the Ni-Fe and Co-Fe magnetizations is ensured. In order to better understand these results, x-ray absorption measurements in total electron yield (TEY) where done on samples prepared only up to the Al_2O_3 barrier. Here, we used an Al wedge with thickness ranging from 0.5 nm to 5 nm and oxidation conditions optimized for 1.2 nm. Thus there are heavily over- and underoxidized regions on the sample:

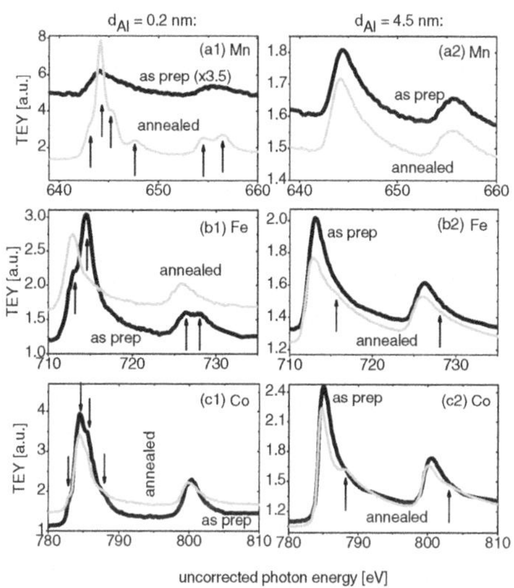

Fig. 3. X-ray absorption signal at the Mn-, Co- and Fe-$L_{2,3}$ absorption edges of the Al-wedge sample in the as-prepared state and after in-situ annealing at 280°C for extreme overoxidation [$d_{Al} = 0.2$ nm, left column] and underoxidation [$d_{Al} = 4.5$ nm, right column]. The measured TEY signal is normalized to the intensity at $h\nu = 500$ eV

The chemical states of the as-prepared sample and after in-situ annealing at 280°C are discussed first. Typical XAS spectra for strongly overoxidized and underoxidized sample regions are shown in Fig. 3. In the as-prepared state the shape of the Mn-$L_{2,3}$ absorption edge (Figs. 3(a1) and (a2)) indicates the metallic state of Mn for all Al thicknesses [20, 21]. The Mn signal is in this state relatively weak, because it originates from the buried Mn-Ir layer. After in-situ annealing two different edge shapes are found; for high Al thickness the Mn is still in the metallic state, but with decreasing Al thickness a Mn multiplet structure characteristic for MnO_X appears [22]. Simultanously, the Mn absorption line intensity increases indicating that Mn diffuses from the buried AFM IrMn towards the barrier. In the as-prepared state the Fe is in the metallic state only for high Al thicknesses which guarantees an underoxidation of the sample, see Figs. 3(b1) and (b2). For smaller d_{Al} the line shape changes from the broad metallic peaks to a multiplet characteristic for FeO_X [23]. After annealing at 280°C the FeO_X signature disappears even for very thin Al layers as in Fig. 3(b1). Because simultaneously, a MnO_X fingerprint appears, the FeO_X obviously is reduced by the diffusing Mn which is then oxidized instead of the Fe. A simple explanation for this behaviour is found by analyzing the enthalpies of oxide formation of Fe and Mn; they are larger for the Mn-oxides than for the corresponding Fe-oxides. From the thermodynamic point of view, the formation of MnO_X both reduces the FeO_X as well as acts as a sink for diffusing Mn. This is in agreement with the observation of an increasing Mn intensity at low Al thickness (equivalent to an increased overoxidation of the samples) which goes along with the more pronounced MnO_X formation. As suggested by Parkin et al. [24] the reduction of the lower electrode in overoxidized samples is important with respect to the thermally induced TMR increase. Similar to the Fe line the shape of the Co-$L_{2,3}$ absorption edge is changed with decreasing Al thickness, see Figs. 3(c1)–(c2). Starting from the typical structurless metallic line shape at high Al thickness in the as-prepared state, the Co-L_3 absorption edge reveals an additional contribution of the Co-oxid multiplet at very low Al thickness. But in contrast to Fe, the Co multiplet remains after annealing, indicating that the CoO_X is not significantly reduced by the diffusing Mn, although the enthalpy of formation for the individual Co-oxides is again smaller than for corresponding Fe- and Mn-oxides. This difference is reasonable, because the oxidation of Co-Fe grains results in a segregation of Fe to the surface of the grain (this is revealed by Auger depth profiling of plasma oxidized Co-Fe single layers). Furthermore, at moderate temperature ($T_A \leq 400$°C) grain boundaries are the prominent diffusion paths in our polycrystalline samples with a typical grain size of only a few 10 nm. Therefore, we may assume that FeO_X at the grain surface can react with the Mn diffusing along the grain boundaries and that the CoO_X should be somewhat buried inside the grain. The interjacent (Fe-Mn)O_X hinders its reduction process requiring a transfer of oxygen from Co to Mn. For strongly underoxidized sample regions the Co-$L_{2,3}$ absorption edge (Fig. 3(c2)) shows additional shoulders about 3 eV above the white lines after annealing. This is also found for the Fe-$L_{2,3}$ absorption edge (Fig. 3(b2)), but less pronounced. This spectral feature indicates the formation of a Co-Fe-Al alloy driven by an interdiffusion of the residual metallic Al into the Co-Fe electrode [25].

Now the thermally induced alterations of the chemical states at the Co-Fe / AlO_X interface of half MTJs, which represent the oxidation range from moderate under- to moderate overoxidation of the Al layer, are discussed for annealing temperature of T_A = 175°C, 275°C and 350°C. Here, instead of the Al wedge, we used a sample with an Al thickness of 1.4 nm, where the optimum oxidation time is 100s. As expected, the maximum intensity of the O-K absorption edge at $h\nu$ = 540.5 eV increases with longer oxidation time

(Fig. 4). These relations are preserved after annealing. The shape of the Fe-$L_{2,3}$ absorption edge in the as-prepared state for 10s oxidation corresponds to metallic Fe, whereas for 999s oxidation the superimposed multiplet structure indicates the partial formation of FeO_X. For optimal oxidation (100s) the multiplet structure is not clearly visible, but a small shoulder on the high energy side of the white lines indicates a small amount of FeO_X. Although Fe is partially oxidized for t_{Ox} = 999s, XAS gives no hints to CoO_X in this sample reflecting again the preferential oxidation of Fe. As for the Al-wedge stack, Mn is in the metallic state for all oxidation times in the as-prepared state. The FeO_X component disappears after annealing at T = 275°C. In parallel, the Mn intensity is strongly enhanced with increasing temperature and oxidation time (Fig. 4), but the MnO_X multiplet is only found for oxidation times of 100 s / 999 s and annealing temperature T_A = 275°C.

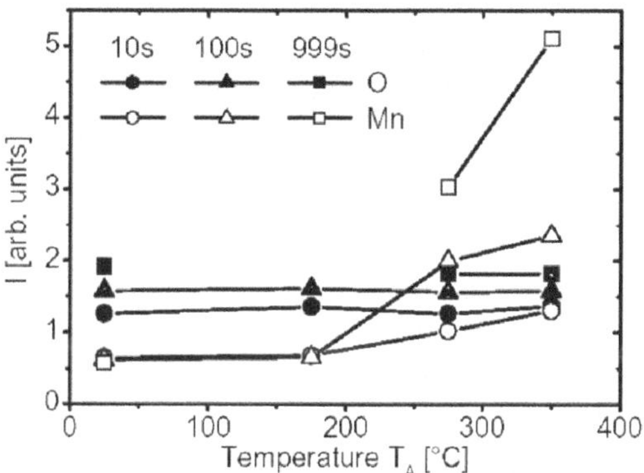

Fig. 4. Maximum TEY-signal I of the O-K and the Mn-L_3 absorption edge of half MTJ stacks for different oxidation times and annealing temperatures

Obviously, the existence of FeO_X prior to annealing stimulates the Mn diffusion and is a prerequisite for the MnO_X formation. Note, that the enthalpy of formation for AlO_X is larger than for MnO_X. For t_{Ox} = 10s XAS gives no hints that the Mn diffusing to the barrier is in an oxidic state. Therefore, at least for the underoxidized samples a diffusion of Mn into the AlO_X is not observed. For longer oxidation time a mixing of MnO_X and AlO_X cannot be ruled out. Auger depth profiling of the full and the half MTJ stacks reveals that the diffusing Mn is preferentially located directly below the barrier after annealing at T_A = 275°C. Mn diffuses from the Mn-Ir towards the barrier and the maximum Mn concentration in the region of the original AFM layer is reduced. At each single annealing temperature, the Mn diffusion towards the barrier becomes stronger with longer oxidation time. Similar results have been found by Lee et al. [26] but in their discussions, the influence of different diffusion paths (grain boundary versus bulk diffusion) was not explicitly taken into account. The shape of the Co-$L_{3,2}$ absorption edge is not significantly altered by annealing; as in the as-prepared state it corresponds to the metallic state for all t_{Ox}. For $t_{Ox} = 10$ s, the amount of residual metallic Al is too small to produce significant changes of the Co- and Fe-XAS spectra. The element-specific magnetic properties of the half MTJ stacks are investigated by XMCD. The relative XMCD signal A_{total} is defined as:

$$
\begin{aligned}
A_{total} &= \frac{-6p+8/3)q}{rP_{h\nu}\cos\theta} \\
with \quad r &= \int_{L3+L2} (I^+ - I^- - a.f.)dE \\
p &= \int_{L3} (I^+ - I^-)dE \\
q &= \int_{L3+L2} (I^+ - I^-)dE
\end{aligned}
$$

For parallel (antiparallel) alignment of the photon spin and the projection of the magnetization on the X-ray propagation direction the corresponding TEY spectrum is denoted I^+ (I^-). The background function a.f. is a two-stepfunction with thresholds set to the peak positions of the L3 and L2 resonances and relative step heights of 2/3 (L3) and 1/3 (L2). The XMCD asymmetries defined as $(I^+ - I^-)$ for Co and Fe are found to be independent of the oxidation time and annealing procedure, e.g., the ratio of the orbital to spin magnetic moments is constant for Fe and Co. Especially the partial FeO_X formation resulting in a changed XAS line shape does not change the shape of the XMCD asymmetry and, therefore, the magnetic moments of the FeO_X are not ferromagnetically ordered. The relative XMCD signals A_{total} of Fe and Co are summarized in Fig. 5.

For underoxidized samples (10s) A_{total} of Fe does not show a significant temperature dependence, the magnetic moment of Fe is not changed. For optimal oxidation (100s) A_{total}(Fe) increases after annealing at temperature = 275°C, which corresponds to an increased magnetic moment for Fe. This increase is strong for the overoxidized sample, but also for the highest annealing temperature of 350°C A_{total}(Fe) does not reach the corresponding values of the milder oxidized samples. This behaviour indicates that the annealing temperature dependence of A_{total}(Fe) is directly correlated with the reduction of FeO_X. The development of the relative XMCD signal is different for Co: For all oxidation times A_{total}(Co) is nearly the same up to 275°C, but after annealing at 350°C it is reduced for all samples. Furthermore, A_{total}(Co) depends on the oxidation state of the samples; it is maximum (minimum) for the optimal oxidized (overoxidized) sample. Although the reason is not clear up to now it has to be stressed, that the measured magnetic moment of Co is largest throughout the complete annealing temperature range for the optimal oxidized MTJs. The XMCD asymmetry of Mn nearly vanishes for all samples, although the intensity and the shape of the Mn-$L_{2,3}$ absorption edge

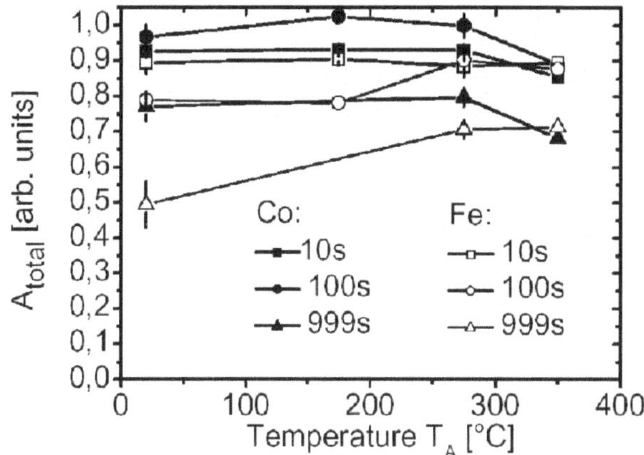

Fig. 5. Relative XMCD signal A_{total} of half MTJ stacks for t_{Ox} = 10 s, 100 s and 999 s and different annealing temperature

depends strongly on the oxidation time and the annealing procedure. The residual very small Mn XMCD signal results from uncompensated Mn spin at the Mn-Ir / Co-Fe interface [27].

1.2 Geometry

Junctions with different shapes and sizes were investigated: rectangular junctions ranging from 700 nm × 700 nm to 700 nm × 1400 nm and elliptical patterns with 500 nm short axes and 850 nm long axes. The patterns were covered by a 15 nm thick Ta layer which minimizes stray field effects of the homemade, CoCr covered MFM tip and, hence, tip induced perturbations of the soft layer magnetization. Sufficient signal to noise ratio and small perturbations were obtained for a CoCr thickness of 30 nm. For the MFM investigations, a modified Nanoscope III from Digital Instruments was operated in the Lift-ModeTM. The magnetic field was generated by two pairs of coils surrounding the microscope. MFM images of the magnetization of the patterned NiFe soft electrodes were recorded at different external fields. As an example, we discuss here the results obtained both with MFM measurements and with corresponding micromagnetic numerical simulations [28] for elliptical junctions [8]. Typical examples are shown in Fig. 6.

In the calculations for elliptical patterns a common feature is found (Fig. 6). Elliptically patterned electrodes often show a high remanent magnetization [M_X/M_S = 0,98, Fig. 6(a),(c)]. The shape, however, favors vortex formation due to minimization of the stray field energy. Consequently, the magnetization reversal of elliptical junctions is often dominated by vortex nucleation and vortex motion with high saturation fields [Fig. 6(b)–(c)]. The results of the simulations are experimentally proven by MFM investigations

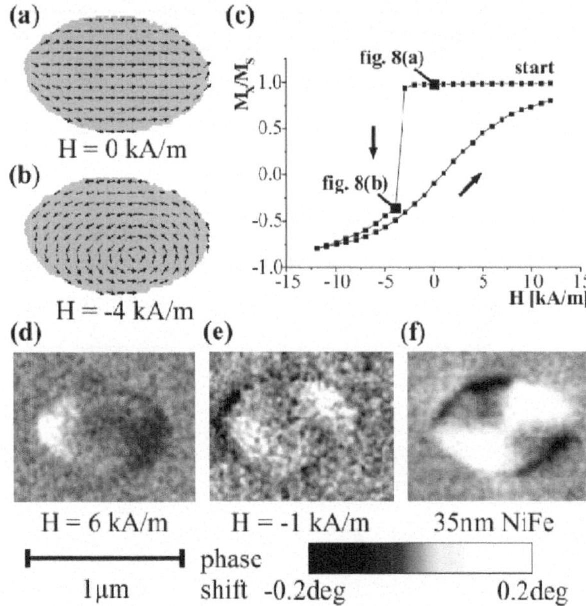

Fig. 6. Vortex nucleation in 500 nm × 850 nm elliptical junctions. (a)–(b) Typical magnetization configurations. (c) Related magnetization curve. (d)–(e) MFM images of elliptically patterned magnetic tunnel junctions recorded at different stages of the minor loop. Additionally an experimental MFM image of a vortex state in a 35 nm thick NiFe ellipse is shown (f)

of the elliptically shaped MTJs, where the complete layer stack including the antiferromagnet was patterned. In saturation or near saturation the NiFe electrodes show a high magnetic contrast at their end points [Fig. 6(d)]. At $H = -1$ kA/m the magnetization shows four opposite regions with bright or dark contrast, which is typical for a vortex state [Fig. 6(e)]. Thicker films with higher contrast show similar patterns more pronounced due to the larger stray fields [Fig. 6(f)]. Here, it should be noted, that considerably different magnetic behaviour was found for nominally identical shapes. This could be traced back to the individually shaped edges of the patterns which result from the grainy structure of the films. Frequently, domain wall pinning was found at kinks or bumps at the edge as small as around 10 nm. MTJs smaller than around 300 nm did not produce enough signal for a reliable MFM measurement. We therefore customized a commercial AFM for electrical measurements with a diamond coated conducting tip. With this instrument, we were able to take TMR minor loops of junctions with sizes down to around 50 nm [6].

Whereas on rectangular and elliptical MTJs results similar those of the MFM investigations were obtained for sizes down to abut 100 nm, truncated

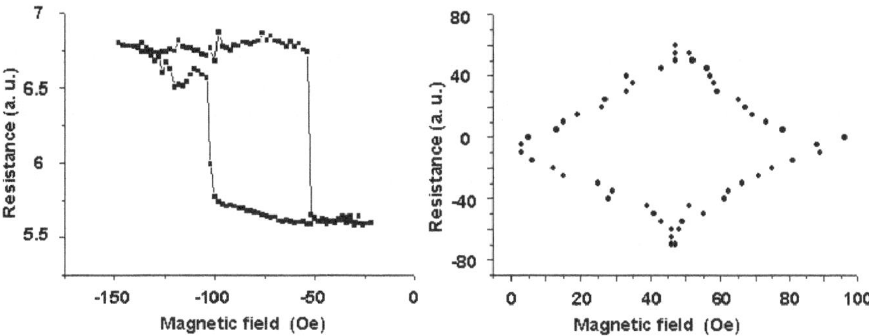

Fig. 7. A single minor loop for a circular MTJ with a diameter of 50 nm (left) and Astroid of a truncated elliptically shaped MTJ with a reproducible switching behaviour (right)

elliptically shaped patterns turned out to show a reproducible switching behaviour as illustrated in Fig. 7 by an astroid for a nominally 200 nm wide MTJ. On junctions with sizes well below 100 nm we were up to now only able to take minor loops.

In Fig. 7, a typical result for a 50 nm circular MTJ is shown. For these ultrasmall elements, we never observed steps in the minor loops or unusually large saturation fields which would point to domain switching or vortex formation, respectively. This and the shape of the minor loops (Fig. 7) therefore suggest a single domain behaviour at these small sizes, which could be a considerable advantage regarding downscaling issues in MRAM applications.

2 New Materials

Introducing new materials in MTJs has two main objectives:

First, the ferromagnetic electrodes can be tailored for giving a large TMR ratio and / or for obtaining a certain dependence of the junction resistance on the external magnetic field. Second, the insulating barrier can be changed; here, the main issues are again a high TMR ratio und in most cases a low area resistance product of the barrier in order to guarantee scalability of the TMR junctions. First, we discuss an example of new electrode materials:

2.1 Heusler Alloys

One possibility to further increase the TMR values, which is frequently discussed, is the use of highly spin polarized materials. Heusler alloys with a predicted gap at the fermi level for only one spin direction are thus very promising candidates for this effort. We prepared thin films of the (so called full)

Heusler alloy Co_2MnSi which is theoretically predicted to have a magnetic moment of around $5\mu_B$ per unit cell at room temperature and a gap for one spin direction of around 0.4 eV [10]. These values critically depend on the degree of ordering of especially Mn and Co and should decrease with increasing disorder. On thin films on a V buffer, we obtained [29] a maximum magnetic moment of $4.7\mu_B$ per stoichiometric unit after annealing at around 400°C. In Fig. 8 we show the dependence of the saturation magnetic moment Ms and the resistivity ρ on the annealing temperature.

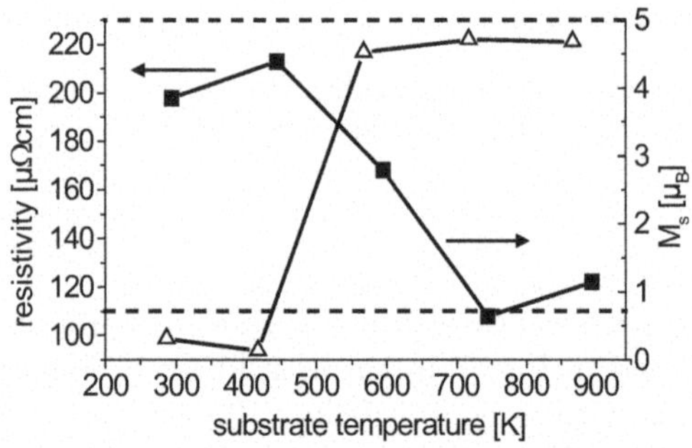

Fig. 8. Dependence of the saturation magnetic moment M_S and the resistivity ρ on the annealing temperature for the Heusler alloy Co_2MnSi

Simultaneously to the maximum of M_S a minimum in ρ occurs, i.e. both properties give evidence for a maximum of the structural order at 400°C. This relatively low annealing temperature is related with the presence of the V buffer layer, which already induces a texture in the Heusler film during the growth at room temperature. These films are then integrated in tunnelling elements with a stacking sequence of V / Co_2MnSi / Al_2O_3 / CoFe / MnIr / Cu / Ta / Au. The problem in producing these elements is the need of a relatively high temperature for ordering the Heusler alloy and only about 275°C for inducing the exchange bias by the IrMn and improving the barrier properties. Thus the first annealing step at high temperature was done after the oxidation of the barrier, the second one after the complete stack was prepared. AFM imaging of the topography showed a rms roughness of about 0.2 nm, i.e. the moderate annealing did not lead to an increase of the surface roughness.

Figure 9 shows first results of a magnetization measurement of this stack and resistivity characteristic. Although the magnetization shows a distinct and separated switching of the pinned and the Heusler-electrode, the dj/dU vs. voltage characteristic is not yet at optimum. This is shown by evaluating the I/V curve of Fig. 9 with a Brinkmann-fit [30], giving the barrier parameters of a barrier height of only 1.6 eV, a large barrier asymmetry of 0.65 eV and an incorrect barrier thickness. This gives evidence either for a not op-

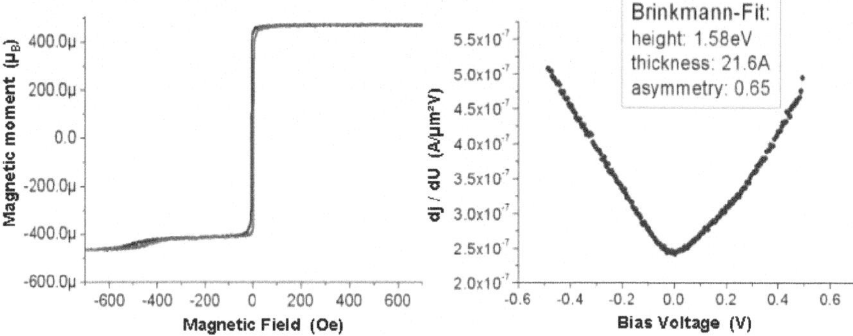

Fig. 9. Magnetization curve of a tunnelling element with Co_2MnSi as soft electrode (left) and the corresponding dependence of dj/dU on the bias voltage (right)

timally oxidized Al or an interlayer between barrier and ferromagnet, both of which usually suppress the TMR signal. In order to clarify this question, TMR measurements at different temperatures and bias voltages have been done.

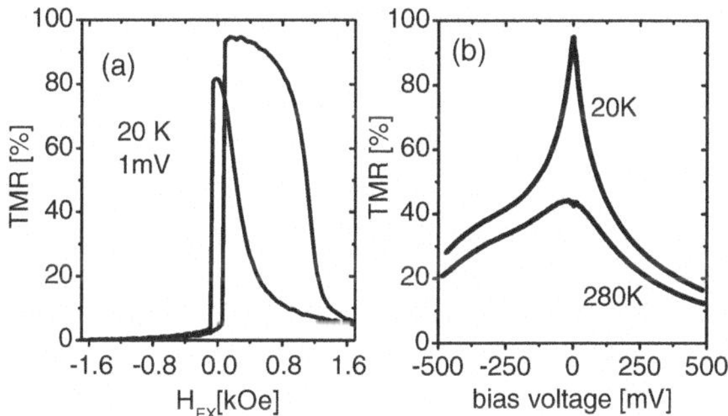

Fig. 10. Dependence of the resistance on an external magnetic field for the tunnelling element with the Heusler alloy Co_2MnSi as soft electrode for different tunnelling areas, temperatures and bias voltages

Figure 10 shows a major loop of an junction with Co_2MnSi electrode and its bias voltage and temperature dependence. At very small bias voltage and low temperature, the major loop TMR reaches nearly 100% again indicating the potential of this material. The strong dependence on both the temperature as well as on the bias voltage again suggests either enhanced scattering of the electrons at impurities or magnons. In order to clarify the reasons, again X-ray absorption measurements of these tunnelling stacks are shown in Fig. 11.

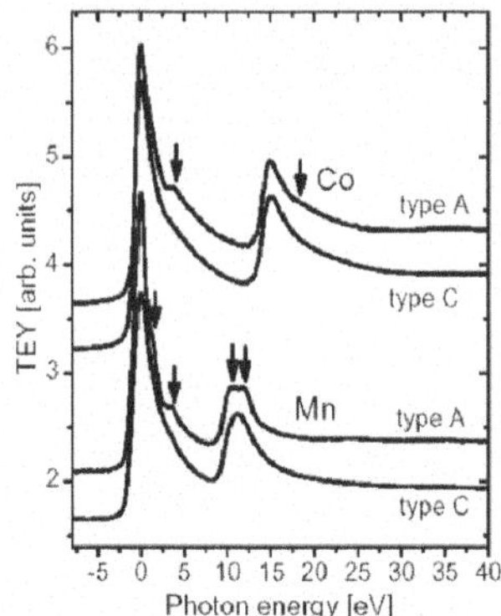

Fig. 11. Typical XAS spectra of a sample V 42 nm / $Co_2MnSi_{d(CMS)}$ / Al 1.4 nm + 200s plasma oxidation + in-situ annealing (A) and a sample V 42 nm / $Co_2MnSi_{d(CMS)}$ / Al 1.4 nm + natural oxidation (B) measured at normal incidence. The photon energy is defined with respect to the maximum of the L_3-absorption edge. The arrows indicate additional features in the XAS of type A samples in comparison to the type B sample. The $L_{2,3}$-resonances are also called white lines

Here, we compare samples which have been not intentionally oxidized (i.e. have the native oxide of Al) with samples from our ECR oxidation procedure. Again, a signature of MnO_X can be found in these samples, where now the Mn originates directly from the (soft) Heusler electrode. This finding thus suggests that efforts must be made in order to avoid Mn or to suppress its diffusion towards the barrier in order to enhance the temperature and bias voltage stability of the TMR in this system. Experiments concerning these issues are under way.

As mentioned, the main other posssibility for developing optimized MTJs is to integrate new barrier materials instead of the commonly used Al_2O_3.

2.2 MgO Barriers

In the last few years, MTJs with MgO barrier have attracted increasing attention due to the theoretically predicted huge TMR effect [12]. It turned out, it is also very favorable in applications because both a low resistance and high thermal stability can be achieved as well [13, 14].

In contrast with the amorphous Al_2O_3 barriers, MgO turned out to be crystalline after an annealing step between 300°C and 400°C, giving rise to a coherent tunnelling of electrons and thus enabling a much larger TMR effect. In Fig. 12, we show a TMR minor loop of a tunnelling junction with an MgO barrier optimized for low resistance and the corresponding dI/dU curves for parallel and antiparallel alignement of the magnetizations:

As can be seen from Fig. 12, a TMR effect of 130% can be achieved with the combination of CoFeB electrodes with MgO barriers even at an

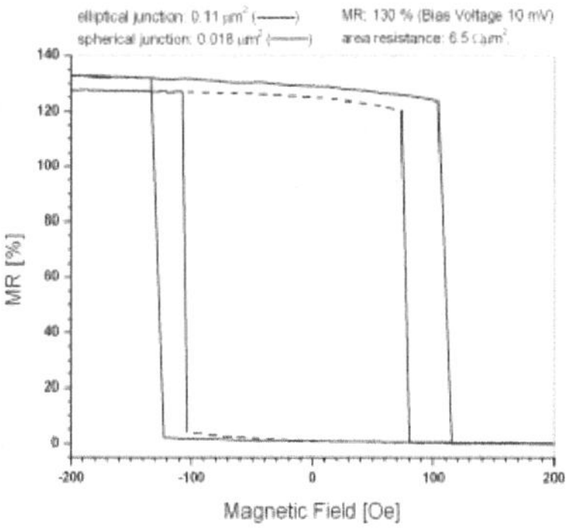

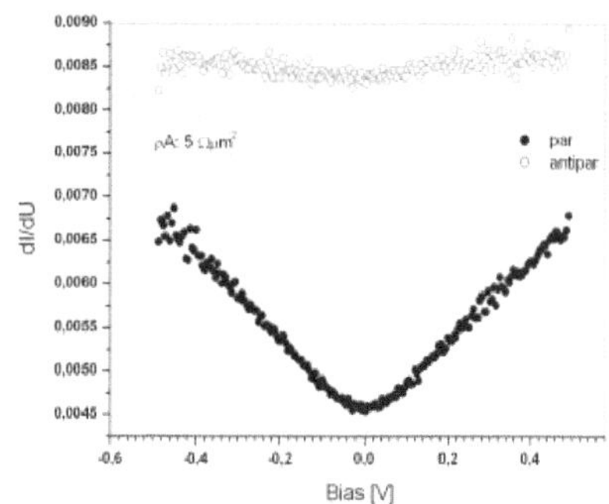

Fig. 12. Resistance vs magnetic field of a MTJ CoFeB / MgO / CoFeB for different shapes (left) and the differential conductance in the parallel and antiparallel state (right)

area resistance product as small as 6.5Ω. This opens scalability of the MTJs down to the sub-100 nm regime, which is necessary for the application in modern electronics.

One striking feature of these junctions is their similarity with Al_2O_3 based MTJs in almost any respect (e.g. temperature and bias dependence of the TMR) except the spectroscopic differential tunnelling conductance shown also in Fig. 12. In the antiparallel state, the conductance is nearly ohmic, meaning a constant value of dI/dU over the entire voltage range, whereas for parallel alignement a combination of quadratic and linear dependence of dI/dU on U occurs as it was also seen for Al_2O_3.

Thus these MgO based junctions need further investigation concerning their unusual transport and tunnelling properties.

3 Applications Beyond MRAM: Magnetic Logic

TMR-effects can have considerably broader field of applications than only as storage cells for MRAMs. Processing units from magnetic tunnelling cells could, e.g., considerably contribute to create a complete technology platform based on magnetic tunnelling junctions. Thus, we concentrate on the use in field programmable logic gate arrays, which is most closely related to the MRAM development in respect mainly to ASICs and embedded memories in logic chips.

Field programmable means that the logic function of a gate array can be changed during the operation of the processing unit. Up to now, this is done by SRAM and FLASH combination, were programming is time consuming and requires relatively large voltage in an, e.g., floating gate architecture.

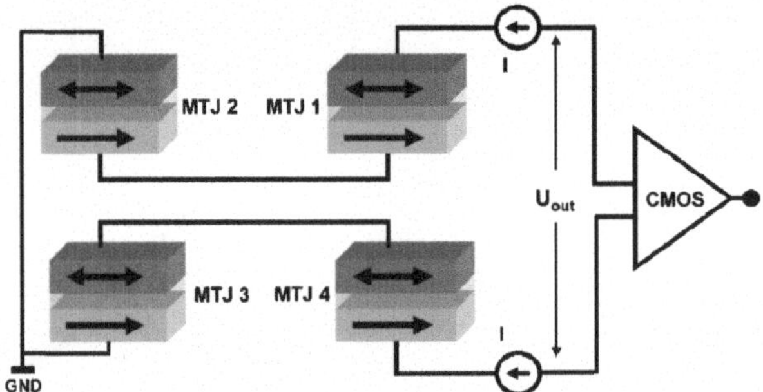

Fig. 13. Bridge configuration of MTJs. The input/programming lines produce a magnetic field that rotates the soft magnetic electrode's magnetization, changing the output voltage V_{out}, which represents the logic function of the inputs

Using a bridge configuration of TMR cells as shown in Fig. 13 could overcome these drawbacks.

Here, the input is represented by currents on two (of four) input lines, which can change the magnetization state of the two adjacent MTJs soft electrodes. Two neighbouring lines are used to set the resistance states of the other two MTJs which 'programme', i.e. define the value V_{out} obtained as logic function of the two inputs.

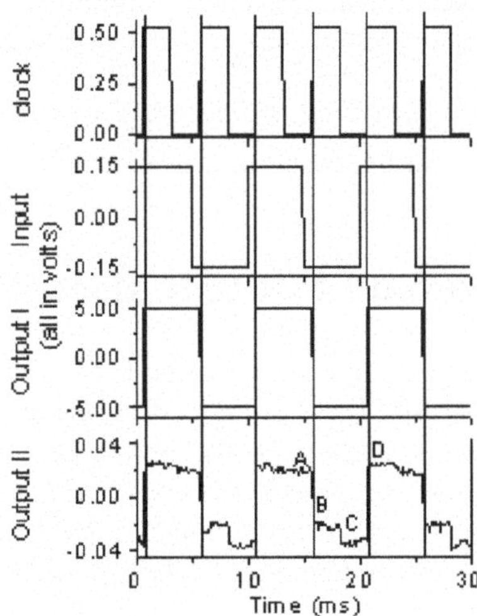

Fig. 14. Clocked operation of six MTJs in a bridge arrangement of sic MTJs. Output 1 is a rectified signal of the original V_{out} (output II).

In Fig. 14, we show a clocked operation of such an arrangement with a programmed “AND” function [15]. This feasibility study thus demonstrates, that MTJs can be used in logic gate arrays which are porgrammable “on the fly”. The large advantages of this “Magnetic Logic” are scalability (similar to MRAM), and speed. Programming these arrays will be as fast as the logic operation with typical time scales down to the nsec regime. This opens in turn new perspectives for innovative schemes like reconfigurable computing [31].

4 Summary

In this contribution, we have discussed aspects of stacking sequence and geometry on the properties of magnetic tunnelling junctions. New materials as, e.g., Heusler alloys can be integrated into the standard stacking sequence of the most advanced MTJs, thereby allowing to reliably test the amount of TMR obtainable. Although the very large spinpolarization of Heusler alloys predicted theoretically has not yet been seen in 'real' MTJs, promising results have been already obtained and further improvements seem to be straightforward. Concerning the influence of the geometry, subtle effects of edge roughness on the magnetization switching seem to be a critical point, although certain shapes have been successfully designed for a single domain like switching. At sizes below 100nm, no signs of domain splitting or vortex formation have been seen up to now, which is very promising for the further downscaling of MTJ storage devices. Beyond this application in MRAMs (or read heads for disk drives), MTJs can be used for realizing a field programmable magnetic logic, where programming is as fast as the logic operation itself, opening thereby the fascinating field of reconfigurable computing. Moreover, MTJs are able to not only detect bits on hard disks but also magnetic micro- and nanoparticles which are already in use for biotechnological and medical applications [32]. Thus a possible production of MRAM chips could boost much more possible applications still ahead.

Acknowledgement

Financial support by the Deutsche Forschungsgemeinschaft (DFG) and the Bundesministerium für Bildung und Forschung (BMBF) is gratefully acknowledged. The Advanced Light Source is supported by the U.S. Department of Energy under Contract No. DE-AC03-76SF00098. We also acknowledge a fruitful cooperation with the Singulus Technologies AG.

References

1. J. S. Moodera, L. R. Kinder, T. M. Wong, and R. Meservey, Phys. Rev. Lett. **74**, 3273 (1995)

2. T. Miyazaki and N. Tezuka, J. Magn. Magn. Mater. **74**, 231 (1995)
3. T. Miyazaki and N. Tezuka, J. Magn. Magn. Mater. 139, 231 (1995) 3 S. S. P. Parkin, K. P. Roche, M. G. Samant, P. M. Rice, R. B. Beyers, R. E. Scheuerlein, E. J. O'Sullivan, S. L. Brown, J. Bucchigano, D. W. Abraham, Y. Lu, M. Rooks, P. L. Trouilloud, R.A. Wanner and W. J. Gallagher, J. Appl. Phys. **85**, 5828 (1999)
4. G. Reiss, H. Brückl, A. Hütten, J. Schmalhorst, M. Justus, A. Thomas, S. Heitmann, phys. stat. sol. (b) **236**, 289 (2003)
5. U. K. Klostermann, R. Kinder, G. Bayreuther, M. Rührig, G. Rupp, J. Wecker, J. Magn. Magn. Mat. **240**, 305 (2002)
6. H. Kubota, G. Reiss, H. Brückl, W. Schepper, J. Wecker, Jpn. J. Appl. Phys. **41**, L180 (2002)
7. Y. Lu, R.A. Altman, A. Marley, S.A. Rishton, P.L. Trouilloud, Appl. Phys. Lett. **70**, 2610 (1997)
8. D. Meyners, H. Brückl, G. Reiss, J. Appl. Phys. **93**, 2676 (2003)
9. J. Schmalhorst, M. Sacher, A. Thomas, H.Brückl and G. Reiss, K. Starke, J. Appl. Phys. **97**, 123711 (2005)
10. S. Ishida, T. Masaki, S. Fujii, S. Asano, Physica B **245**, 1 (1998)
11. J. Schmalhorst, S. Kämmerer, G. Reiss, A. Hütten, Appl. Phys. Lett. **86**, 052501 (2005)
12. W. H. Butler, X. G. Zhang, T. C. Schulthess, and J. M. MacLaren, Phys. Rev. B **63**, 054416 (2001)
13. S. S. P. Parkin, C. Kaiser, A. Panchula, P. M. Rice, B. Hughes, M. Samant and S. H. Yang, Nature Mater. **3**, 862 (2004)
14. S. Yuasa, T. Nagahama, A. Fukushima, Y. Suzuki, and K. Ando, Nature Mater. **3**, 868 (2004)
15. G. Reiss, D. Meyners, Reliability of field programmable magnetic logic gate arrays, Appl. Phys. Lett. **88** (2006) 043505
16. S. Gider, B.U. Runge, A.C. Marley, S.S.P. Parkin, Science **281**, 797 (1998)
17. J. Schmalhorst, H. Brückl, G. Reiss, R. Kinder, G. Gieres, J. Wecker, **77**, 3456 (2000)
18. J. Schmalhorst, H. Brückl, G. Reiss, G. Gieres, M. Vieth and J. Wecker, J. Appl. Phys. **87**, 5191 (2000)
19. J. Schmalhorst, H. Brückl, G. Reiss, G. Gieres, and J. Wecker, J. Appl. Phys. **91**, 6617 (2002)
20. S. Andrieu, E. Foy, H. Fischer, M. Alnot, F. Chevrier, G. Krill, and M. Piecuch, Phys. Rev. B **58**, 8210 (1998)
21. B. T. Thole, R. D. Cowan, G. A. Sawatzky, J. Fink, and J. C. Fuggle, Phys. Rev. B **31**, 6856 (1985)
22. L. Stichauer, A. Mirone, S. Turchini, T. Prosperi, S. Zennaro, N. Zema, F. Lama, R. Pontin, Z. Simsa, P. Tailhades, et al., Phys. Rev. Lett. **91**, 017203 (2003)
23. T. J. Regan, H. Ohldag, C. Stamm, F. Nolting, J. Lüning, J. Stöhr, and R. L. White, Phys. Rev. B **64**, 214422 (2001)
24. L. Seve, W. Zhu, B. Sinkovic, J. W. Freeland, I. Coulthard, W. J. Antel jr, and S. S. P. Parkin, Europhys. Lett. **55**, 439 (2001)
25. G. A. Botton, G. Y. Guo, W. M. Temmerman, and C. J. Humphreys, Phys. Rev. B **54**, 1682 (1996)
26. C. S. Yoon, J. H. Lee, H. D. Jeong, C. K. Kim, J. H. Yuh, and R. Haasch, Appl. Phys. Lett. **80**, 3976 (2002)

27. H. Ohldag, A. Scholl, F. Nolting, E. Arenholz, S. Maat, A. T. Young, M. Carey, and J. Stöhr, Phys. Rev. Lett. **91**, 017203 (2003)
28. OOMMF program (release1.1), NIST (Gaithersburg, USA), available at http://math.nist.gov/oommf for public use.
29. S. Kämmerer, A. Thomas, A. Hütten, G. Reiss, Appl. Phys. Lett. **85**, 79 (2004)
30. H.Brückl, J.Schmalhorst, G.Reiss, G.Gieres, J.Wecker, Appl. Phys. Lett. **78**, 1113 (2001)
31. W. C. Black, Jr., B. Das, J. Appl. Phys. **87**, 6674 (2000)
32. J. Schotter, P.B. Kamp, A. Becker, A. Pühler, G. Reiss, H. Brückl, Biosensors & Bioelectronics, **19**, 1149 (2004)

Magnetic Anisotropies in Ultra-Thin Iron Films Grown on the Surface-Reconstructed GaAs Substrate

B. Aktaş[1], B. Heinrich[2], G. Woltersdorf[2], R. Urban[2], L.R. Tagirov[2,3], F. Yıldız[1], K. Özdoğan[1], M. Özdemir[1], O. Yalçın[1], and B.Z. Rameev[1]

[1] Gebze Institute of Technology, 41400 Gebze-Kocaeli, Turkey
[2] Simon Fraser University, Burnaby, BC, V5A 1S6, Canada
[3] Kazan State University, 420008 Kazan, Russian Federation

Summary. Magnetic anisotropies of epitaxial ultra-thin iron films grown on the surface reconstructed GaAs substrate have been studied. Ferromagnetic resonance (FMR) technique has been exploited to determine magnetic parameters of the films in the temperature range 4–300 K. The unusual angular dependence of FMR spectra allowed us to build precise model of the magnetic anisotropies of the studied systems. Presence of strong perpendicular anisotropy have been deduced. Switching of the principal anisotropy axes has been observed in the double ferromagnetic-layer sample. It has been attributed to drastic relaxation of the uniaxial component of anisotropy induced by the surface reconstruction of the substrate. The linear variation of magnetic anisotropy parameters with the temperature has been observed and discussed in terms of magneto-elastic anisotropies controlled by thermal expansion coefficients of the materials in a contact.

1 Introduction

The interest to ultra-thin magnetic multilayers has been steadily increasing. It is motivated by the fact that magnetic properties of this type of structures are the real technological issues in mass production of data storage devices and magnetic random access memories. A good grasp of the fundamental physics of the magnetization dynamics becomes of essential importance to sustain the exponential growth of device performance factors.

The magnetic anisotropy of the thin films is of crucial importance in applications. It is well known that ferromagnetic resonance (FMR) is the most sensitive and accurate technique to determine magnetic anisotropy fields of very thin magnetic films [1, 2]. In this paper we study the magnetic anisotropies in single and double iron layer structures grown on the surface-reconstructed GaAs single-crystalline substrate and demonstrate how surface-induced anisotropy can be used to tailor overall magnetic properties of the studied system. In our experiments we observed unconventional triple-mode FMR spectra. They are interpreted and explained based on the model proposed in this study. Consistent fitting of angular and frequency behaviors of the FMR spectra in the temperature range 4–300 K allowed us to determine accurately the cubic, uniaxial and perpendicular components of the magnetic anisotropy, as well as establish directions of easy and hard axes for the magnetization in the layer(s). The origin and temperature dependence of the magnetic anisotropy fields are extensively discussed in terms of the surface-induced anisotropies and thermal expansion coefficients of the materials subject to a stress, induced by lattice mismatch.

2 Experimental Results

2.1 Samples Preparation

The single and double iron-layer ultrathin film structures (Au/Fe/GaAs, Cr/Fe/GaAs, Au/Pd/Fe/GaAs, Au/Fe/Au/Fe/GaAs) were prepared by Molecular Beam Epitaxy (MBE) on (4×6) reconstructed GaAs(001) substrates. A brief description of the sample preparation procedure is as follows. The GaAs(001) single-crystalline wafers were subject to annealing and sputtering cycles and monitored by means of reflection high energy electron diffraction (RHEED) until a well-ordered (4×6) reconstruction appeared [3]. Then GaAs substrates were heated to approximately 500 C in order to desorb contaminants. Residual oxides were removed using a low-energy Ar^+ bombardment (0.6 keV) under grazing incidence. Substrates were rotated around their normal during the sputtering. The (4×6) reconstruction consists of (1×6) and (4×2) domains with the (1×6) domain is As-rich, while the (4×2) domain is Ga-rich.

The Fe films were further deposited directly on the GaAs(001) substrates at room-temperature from a resistively heated piece of Fe at the base pressure of 1×10^{-10} Torr. The film thickness was monitored by a quartz crystal microbalance and by means of RHEED intensity oscillations. The deposition rate was adjusted at about one mono-layer (ML) per minute. The gold layer was evaporated at room temperature at the deposition rate of about one monolayer per minute. RHEED oscillations were visible for up to 30 atomic layers. Noble metals are known to have long spin-diffusion length that makes them suitable as a spacer-layer in the spin-valve magnetic field sensor applications. Films under study were covered by a 20ML thick Au(001) or Cr cap layer for protection in ambient conditions. More details of the sample preparation are given in Ref. [3]. The sketch of the single-layer film, coordinate system and principal vector directions are shown in Fig. 1.

2.2 FMR in the Single-Layer Samples

General Measurement Procedure

Main FMR measurements have been carried out using the commercial Bruker EMX X-band ESR spectrometer equipped by an electromagnet which provides a DC magnetic field up to 22 kG in the horizontal plane. The small amplitude modulation of the field is employed to record field-derivative absorption signal at the temperature range 4–300 K. An Oxford Instruments continuous helium-gas flow cryostat was used to cool the sample down to the measurement temperatures, and the temperature was controlled by the commercial LakeShore 340 temperature-control system. A goniometer was used to rotate the sample around the rod-like sample holder in the cryostat tube. The sample-holder was always perpendicular to the DC magnetic field and parallel to the microwave magnetic field. The samples were placed on the sample-holder in two different geometries. For the in-plane angular studies the films were attached horizontally on the bottom edge of the sample holder. During rotation, normal to the film plane remained parallel to the microwave field, but the external DC magnetic field held different directions in the sample plane. This geometry is not conventional and gives quite asymmetric absorption curves but still at the same resonance field as in the conventional geometry for the in-plane measurements (both DC and microwave magnetic fields always lie in the film plane). Thus, we could be able to study at least the angular dependence of the resonance field and magnetic anisotropies using the FMR data taken from the unconventional in-plane geometry. We have also recorded some FMR data in the conventional, in-plane geometry for some specific crystallographic direction to check correctness of the data obtained in the unconventional geometry.

For the out-of-plane measurements the samples were attached to a flat platform precisionally cut at a cheek of the sample holder. Upon rotation of the sample the microwave component of the field remained always in the sample plane, whereas the DC field was rotated from the sample plane towards the film normal to get additional data for accurate determination of the anisotropy fields.

In-Plane FMR Measurements

First we measured the in-plane FMR in the 20Au/15Fe/GaAs(001) sample. Figure 2a illustrates a temperature evolution of the in-plane FMR spectra taken in the direction of the DC magnetic field $\mathbf{H}\|\mathbf{a}$, where $\mathbf{a}$ is the [110] direction of GaAs substrate or iron film ($\mathbf{a}\|\mathbf{x}$ in Fig. 1). The single, relatively narrow and intensive signal is observed at very low magnetic fields in the entire temperature range. As the temperature decreases starting from room temperature, the resonance field of FMR signal steadily shifts from ~ 320 G for 300 K down to about 150 G at 5 K. The line width of the resonance increases with decreasing the temperature.

Contrary to the measurements along the $\mathbf{a}$ axis, the in-plane FMR spectrum in the $\mathbf{b}$ direction (the $[1\bar{1}0]$ direction of the GaAs substrate) unexpectedly consists of three signals (labeled by P_1, P_2 and P_3, Fig. 2b). As far as we know it is a unique observation of three FMR signals from a single, homogeneous ferromagnetic film. Usually, a single resonance peak (mode) is expected from such a very thin (15 monolayers) ferromagnetic layer, since the spin wave

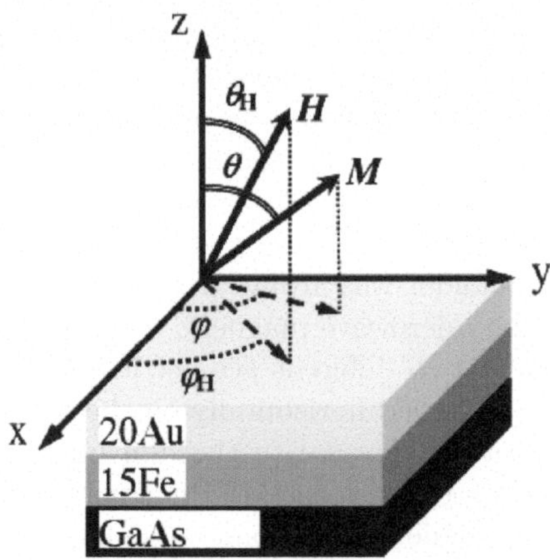

Fig. 1. The sketch of the samples studied in the paper.

modes are expected to shift to negative fields due to very high excitation energies of the short-wavelength standing spin waves across the thickness of the ultrathin film. The temperature dependence of the FMR spectrum for $\mathbf{H}\|\mathbf{b}$ is shown in Fig. 2b. All three peaks are observable in the entire temperature range. The high-field signal has largest intensity at all temperatures. At room temperature the two low-field peaks overlap and almost merge into a common signal of distorted shape. As the temperature decreases, the low-field signal separates into two signals which shift one from the other in opposite directions. The intense high-field signal shifts monotonically to higher fields upon lowering the temperature. At lowest measurement temperatures (4–5 K) the overall splitting of the FMR spectra reaches $\sim$ 1700 G. It should be noted that the high-field mode for $\mathbf{H}\|\mathbf{b}$ shifts in the opposite direction to that for the $\mathbf{H}\|\mathbf{a}$ axis. This implies that the easy direction for magnetization in the sample plane is the **a** axis (and the **b** axis is the hard direction).

To study the in-plane anisotropy of the film, we rotated the DC magnetic field in the plane of the film (**ab**-plane). The room-temperature in-plane angular dependence of the FMR spectra is given in Fig. 3. As it can be seen from the figure the number of absorption peaks is varied with the in-plane rotation angle. The relative intensity of the signals is also angular-dependent. In fact, the FMR spectra show an overall periodicity of 180 degrees. This implies that the system must have at least uniaxial symmetry in the film plane. The unusual splitting of the spectra on the three components allows us to suppose that a considerable cubic anisotropy component is superimposed on the uniaxial anisotropy. We will see later from our simulations of the FMR spectra that it is actually the case.

In order to increase reliability of the FMR spectra interpretation we have also studied the frequency dependence of the in-plane FMR spectra between 9–36 GHz at room temperature using our high-frequency extension modules.

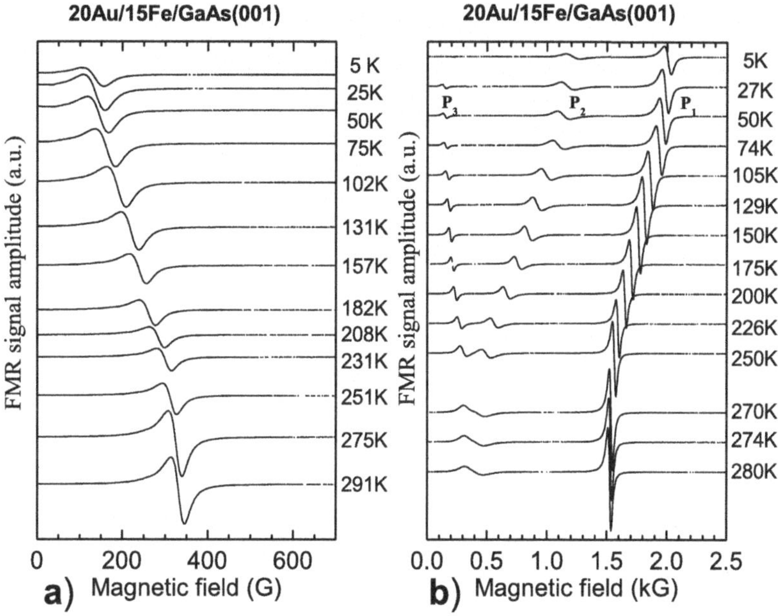

Fig. 2. Temperature dependence of in-plane FMR spectra taken for **H**‖**a** (a), **H**‖**b** (b).

Right-hand-side inset in Fig. 3 shows an angular variation of the in-plane resonance field measured at 24 GHz. Left-hand-side inset shows variation of the resonance fields with the microwave frequency. The parabolic dependence of experimental points on the frequency is clearly seen from the figure. This dependence is a general frequency behavior of the resonance field in systems with strong anisotropy (the anisotropy energy is comparable with the Zeeman energy at low-frequency region).

Out-of-Plane FMR Measurements

We have also made complementary out-of-plane FMR measurements when the DC magnetic field is rotated from the easy, **a**, or the hard, **b**, axes in the film plane towards the normal direction to the film plane. Figure 4 shows the evolution of the FMR field for resonance upon changing the polar angle in the **b**–**c** plane. Double-peak FMR spectra are observed for this geometry. The separation between modes steadily increases with approaching the film normal. As it will be clear later, the second mode does not belong to the higher-order spin-wave modes. However, the angular dependence of resonance field of the modes can be used to clarify the nature of magnetic anisotropy and to obtain accurately the anisotropy parameters from a computer fit.

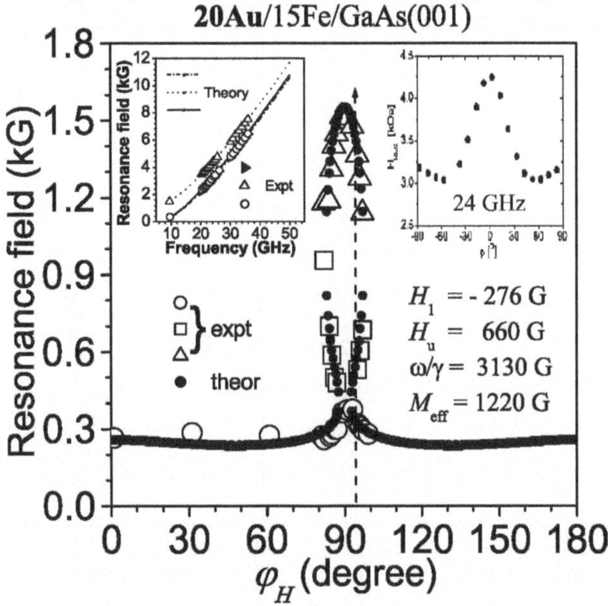

Fig. 3. In-plane (**ab** plane) angular dependence of FMR spectra for the 20Au/15Fe/GaAs(001) sample at room temperature. Right-hand-side inset – the same angular dependence for 24 GHz, left-hand-side inset – the frequency dependence of the resonance field for the hard direction (upper) and easy (lower) directions.

Influence of the Cap Layer Material

We have studied also the sample where the gold cap-layer material had been replaced by chromium: 20Cr/15Fe/GaAs(001). In-plane geometry measurements have shown drastic decrease of the uniaxial component in the in-plane anisotropy of the iron film as presented in Fig. 5a. The picture looks like the influence of the Cr cap-layer results in cancelling of the uniaxial anisotropy induced by the surface reconstruction of the GaAs substrate. At the same time, it is clearly visible that the principal axis of residual uniaxial anisotropy is tilted by about 23 degrees with respect to its original direction for the case of Au-capped sample. Another sample with a composite cap layer, 20Au/9Pd/16Fe/GaAs(001), revealed only minor influence of the palladium interlayer on the magnetic anisotropy of the iron film (see Fig. 5b). However, an additional FMR signal appeared in the main domain of the in-plane rotation angles, which we attribute to the magnetic response of the Pd interlayer (see discussion below).

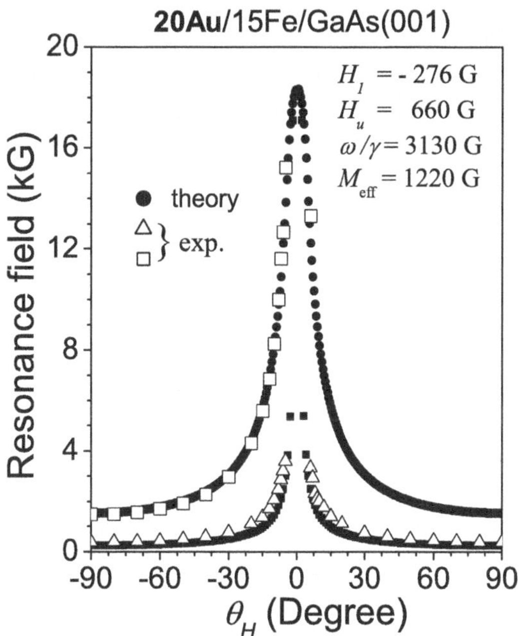

Fig. 4. Out-of-plane angular dependence of resonance field for **H**||**b**.

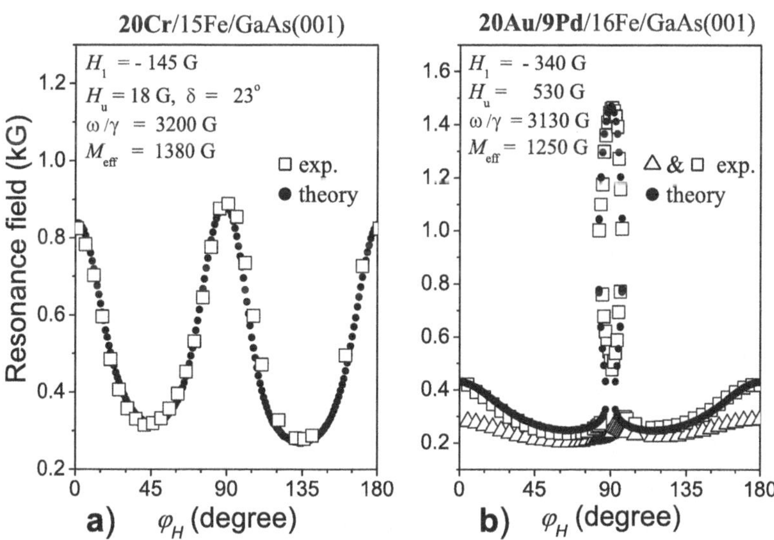

Fig. 5. Influence of the cap layer – in-plane angular dependence of the resonance field: (a) chromium cap layer; (b) composite Pd/Au cap layer.

2.3 FMR in the Double-Layer Samples

After learning the FMR response from the single iron layer we have studied the spin-valve type, double-layer system. Figure 6 shows temperature evolution of the in-plane FMR spectra for the 20Au/40Fe/40Au/15Fe/GaAs(001) sample taken at two angles. We used data of our previous measurements on the single-layer sample, 20Au/15Fe/GaAs(001) as a reference, in which we have established easy and hard directions for the 15ML thick iron layer. The spectra at Figs. 6a and 6b have been recorded at the easy and hard axes for the first, 15ML thick iron layer.

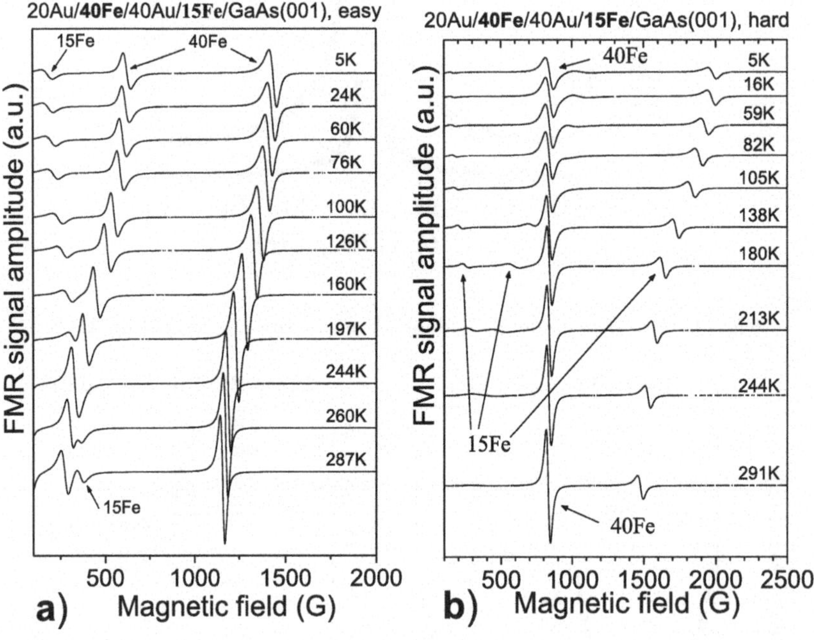

Fig. 6. The in-plane FMR spectra of the double-layer sample for two orientations of magnetic field with respect to the crystallographic axes: (a) DC field is parallel to the hard axis of the first, 15ML-thick iron layer; (b) DC field is parallel to the easy axis of the first layer.

Three FMR absorption peaks are clearly visible in the spectra recorded at the angle labelled "easy". As we do not expect any marked exchange or magnetostatic interaction through the 40ML-thick gold layer, the contribution of the first, 15ML-thick, iron layer to the multi-component FMR signal can be easily attributed (see labelling in the Fig. 6a) by comparison with the measurements on the single-layer sample, Fig. 2. The double-peak signal from the second, 40ML thick, iron layer gives a hint that easy and hard axes of the second iron layer are tilted with respect to the principal anisotropy axes of the first layer.

The FMR spectra recorded in the "hard" direction (Fig. 6b) show four-peak structure at certain range of temperatures, three of which are identified as a hard-axis spectra of the first, 15ML thick, iron layer. Then, the single-peak FMR spectrum of the 40ML-thick layer clearly indicates proximity to the easy direction for this layer.

Full angular dependence of the FMR spectra at room temperature is displayed in Fig. 7 (experimental points are given by empty symbols). From the figure it is evident that the hard axis (angle for maximal resonance field) of the second, 40ML thick, iron layer is switched on the 90 degrees apart of the hard axis of the first, 15ML thick, iron layer. Let us emphasize here the importance of the FMR measurements at different temperatures, Fig. 6, for identifications of the angular dependence of the resonance fields in Fig. 7.

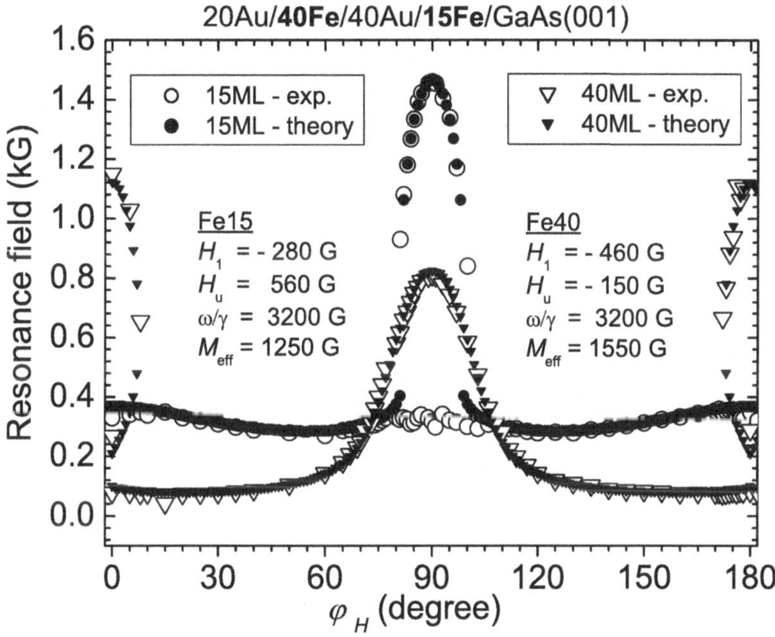

Fig. 7. The angular dependence of the in-plane resonance field for the double-layer sample at room temperature: open symbols – experimental data, solid symbols – results of the fitting.

3 Model and Computer Simulation of the FMR Spectra

3.1 Basic Formulas for the FMR Field for Resonance and Absorption

The FMR data are analyzed using the model free energy expansion in the notations which is similar to that used in Ref. [4]:

$$E_T = -\mathbf{M}\cdot\mathbf{H} + \left(2\pi M_0^2 - K_p\right)\alpha_3^2 + K_1\left(\alpha_1^2\alpha_2^2 + \alpha_2^2\alpha_3^2 + \alpha_3^2\alpha_1^2\right) + K_u\alpha_{[1\bar{1}0]}^2 . \tag{1}$$

Here, the first term is the Zeeman energy in the external DC magnetic field, the second term is the demagnetization energy term including the effective perpendicular anisotropy as well, the third term is the cubic anisotropy energy characterized by the parameter K_1, and the last term is the uniaxial anisotropy energy. In this equation α_i's represent directional cosines [6] of the magnetization vector **M** with respect to the crystallographic axes ([100], [010] and [001]) of GaAs substrate (or iron film), while $\alpha_{[1\bar{1}0]}$ is the directional cosine with respect to $\mathbf{b}\|[1\bar{1}0]$, which is the hard direction. M_0 is the saturation magnetization at the temperature of measurement. It should be remembered here that one of the crystallographic axes is always perpendicular to the sample plane, and the remaining two lie in the sample plane. That is why we could combine demagnetization and perpendicular anisotropy terms in a single term (second one) using only the α_3 directional cosine. The relative orientation of the reference axes, sample sketch and various vectors relevant in the problem are given in Fig. 1. The fields for resonance are obtained using the well known equation [5]:

$$\left(\frac{\omega_0}{\gamma}\right)^2 = \left(\frac{1}{M_0}\frac{\partial^2 E_T}{\partial\theta^2}\right)\left(\frac{1}{M_0\sin^2\theta}\frac{\partial^2 E_T}{\partial\varphi^2}\right) - \left(\frac{1}{M_0\sin\theta}\frac{\partial^2 E_T}{\partial\varphi\partial\theta}\right)^2 , \tag{2}$$

where $\omega_0=2\pi\nu$ is the circular frequency of the ESR spectrometer, γ is the gyro-magnetic ratio for the material of the magnetic film, θ and φ are usual polar and azimuthal angles of the magnetization vector **M** with respect to the reference system. We do not consider standing spin-wave excitations in the film because the film thickness is too small ($\sim$ 20–80 Å).

The imaginary component of the dynamic magnetic susceptibility that corresponds to the absorbed microwave energy by the sample is given by [1, 7]

$$\chi_2 = 4\pi\frac{m_\varphi}{h_\varphi} = \frac{8\pi M_0\omega}{\gamma^2 T_2}\left(\frac{\partial^2 E_T}{\partial\theta^2}\right) . \times\left\{\left[\left(\frac{\omega_0}{\gamma}\right)^2 - \left(\frac{\omega}{\gamma}\right)^2\right]^2 + \frac{4\omega^2}{\gamma^4 T_2^2}\right\}^{-1} \tag{3}$$

Here $\omega = \gamma H$ is the Larmour frequency of the magnetization in the external DC magnetic field, T_2 represents the effective homogeneous relaxation time of the magnetization which contributes to the line width of the FMR signal. We deduce the model parameters as a result of the fitting of the experimental data using the above two equations for computer simulations.

3.2 Analysis of the In-Plane FMR Spectra from the Single-Layer Samples

To analyze the data for the in-plane geometry of measurements both polar angles in Eqs. (1) and (2), θ for magnetization, and θ_H for external DC magnetic field, are fixed at $\theta, \theta_H = \pi/2$, and the azimuthal angle of magnetization φ is obtained from the static equilibrium condition for the directions φ_H of the external field varied in the range from zero to π. Then, the set of equations for the in-plane geometry reads:

$$H \sin(\varphi - \varphi_H) + \frac{1}{2} K_1 \sin 4(\varphi - \varphi_{[100]}) - K_u \sin 2\varphi = 0,$$

$$\begin{aligned} \left(\frac{\omega_0}{\gamma}\right)^2 = & [H \cos(\varphi - \varphi_H) + 4\pi M_{eff} + \frac{1}{2} K_1 (3 + \cos 4(\varphi - \varphi_{[100]})) \\ & -2K_u \cos^2 \varphi] \times [H \cos(\varphi - \varphi_H) + 2K_1 \cos 4(\varphi - \varphi_{[100]}) \\ & -2K_u \cos 2\varphi]. \end{aligned} \quad (4)$$

The effective magnetization M_{eff} includes contribution from the perpendicular anisotropy: $2\pi M_{eff} = 2\pi M_0 - K_p/M_0$. The angle $\varphi_{[100]} = \pi/4$ is the angle between the easy directions of the cubic and uniaxial anisotropies. The results of computer solution for $H = H^{res}_{in-plane}$ of the above system for the 20Au/15Fe/GaAs(001) sample are plotted in Fig. 3 in solid circles. The best-fit parameters are given in the figure. The calculation procedure and Eq. (3) have also been used to simulate the experimental FMR spectra at different temperatures. The simulated spectra are plotted together with the experimental ones in Fig. 8a. There is quite good agreement between the calculated and the experimental angular dependence of the resonance field in Fig. 3 and the temperature evolution of the FMR spectrum in Fig. 8a. Notice here, that the in-plane angular dependence of the resonance field had been fitted simultaneously with the out-of-plane dependence, described below.

The computer simulations revealed that the angular dependence of the field for resonance shows unusual behavior. There is only single resonance line (Fig. 3) in the main domain of the angles. However, when direction of the DC magnetic field approaches the (hard) axis **b** (the angle φ varies $\pm 10°$ around $90°$) double or even triple resonance absorption is observed (see Figs. 2b and 8a). This point can be made more clear if we follow the dash vertical line in the Fig. 3. Moving along the line from zero magnetic field in

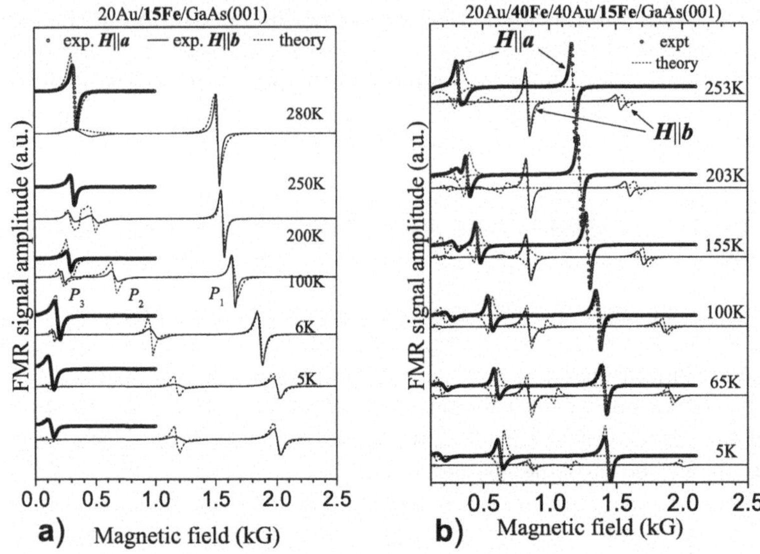

Fig. 8. Simulated FMR spectra taken as some selected temperatures for **H**||**b**: (a) for the single-layer sample; (b) for the double-layer sample.

the upward direction (just like spectrometer sweeps DC magnetic field when recording the FMR spectrum) we observe how the dash line intersects the figure of the angular dependence three times. In the close vicinity of 90° there are two intersections. Every intersection of the dash line with the figure of angular dependence gives FMR signal. For the triple-line FMR spectrum these intersections are marked as P_1, P_2 and P_3 in Figs. 2b and 8a.

In the case of double or triple FMR lines, the lower one(s) corresponds to the non-aligned situation, when the condition for resonance, Eq. (4), is satisfied at the direction of the magnetization, which is not parallel to the external magnetic field. The high-field signal may be called as an aligned signal, when the magnetization and the external DC field are practically parallel. To our best knowledge, no observations of two non-aligned FMR resonances in a single ferromagnetic film had been reported in the literature. The double-peak FMR spectra in a single iron film have been observed previously (see, for example, Refs. [8, 9, 10, 11]). We believe that a specific relation between cubic and uniaxial anisotropies is realized in the studied system, which gives rise the observed unusual behavior of the FMR spectra. However, if the measurement frequency is increased, the angular-dependence loop opens, and the single-line FMR spectrum is observed in the full domain of angles (right-hand-side inset of Fig. 3).

Fitting of the in-plane angular dependence for the Cr-capped sample (see solid symbols in Fig. 5a) revealed drastic suppression of the substrate-induced uniaxial anisotropy by the Cr overlayer. Contrary to this case, the Pd/Au-capped sample shows only minor influence of the palladium interlayer on the anisotropy of the iron layer (see Fig. 5b, solid symbols). However, an additional FMR line appears at low fields which can not be reproduced within our model. We may speculate that this signal comes from the palladium layer magnetized by the proximity with the ferromagnetic iron layer. However, this question needs special investigation.

3.3 Analysis of the Out-of-Plane FMR Spectra from the Single-Layer Samples

For the out-of-plane geometry of measurements the azimuthal angle φ_H is fixed either at $\varphi_H = 0$ (hard axis, i.e. the DC magnetic field is rotated in the (110) plane), or $\varphi_H = \pi/2$ (easy axis, i.e. the DC magnetic field is rotated in the $(1\bar{1}0)$ plane), while the polar angle θ_H is varied from $\pi/2$ to zero. The polar and azimuthal angles of magnetization for each direction of the external field have been obtained from static equilibrium condition corresponding to the minimum free energy of the system. Then, taking the parameter $\varphi_{[100]} = \pi/4$, the set of equations for the out-of-plane measurements from easy axis (**a**) direction can be obtained:

$$H \sin(\theta - \theta_H) - 2\pi M_{eff} \sin 2\theta + \frac{1}{4} K_1 \sin 2\theta \, (3 \cos 2\theta + 1) = 0,$$

$$\left(\frac{\omega_0}{\gamma}\right)^2 = [H \cos(\theta - \theta_H) - 4\pi M_{eff} \cos 2\theta + \frac{1}{2} K_1 (\cos 2\theta + 3 \cos 4\theta)] \times [H \cos(\theta - \theta_H) - 4\pi M_{eff} \cos^2 \theta + \frac{1}{4} K_1 \left(8 \cos 2\theta - 3 \sin^2 2\theta\right) + 2K_u] . \quad (5)$$

Double-peak FMR spectra appeared also for the out-of-plane geometry when measured from the hard in-plane direction. The results of the computer solution of the system Eq. (5) obtained in the procedure of simultaneous fitting with the in-plane dependence from Fig. 3 for the 20Au/15Fe/GaAs(001) sample are plotted in Fig. 4 in solid symbols. The best-fit parameters are given in the figure. Thus, the calculations show fairly good agreement at all angles, in-plane and out-of-plane geometry, all temperatures and frequencies of the measurements. The deduced magnetic parameters used in the calculation of the simulated spectra are given in Fig. 9a.

3.4 Analysis of the In-Plane FMR Spectra from the Double-Layer Sample

The comparison of the spectra from Fig. 2a with the spectra from Fig. 6a, recorded in the easy direction of the first, 15ML-thick iron layer, says us that the double-line FMR spectrum is observed for the second, 40ML-thick iron layer. And *vice-versa*, the spectra in Fig. 6b, recorded in the hard direction reveal the single-line FMR spectrum from the second iron layer. Being guided with the results of simulations for the single-layer sample we realize clearly that there is inclination angle between the easy and hard axes of the first and the second iron layers. The fitting of the full angular dependence of the second layer FMR spectra using the equation set (4) for the signal from each layer (Fig. 7) revealed that the hard axis of the uniaxial anisotropy term in the second, 40ML-thick iron layer, is switched 90 degrees with respect to the hard axis of that in the first, 15ML-thick iron layer. The fitting parameters shown in Fig. 7 allow to conclude that origin of the observed switching is mainly drastic relaxation and change in sign of the uniaxial component of the magnetic anisotropy in the second layer ($\simeq -150$ G as compared with $\simeq 660$ G for the single 15ML-thick layer at room temperature). Comparison of the model calculations with the temperature evolution of the experimental spectra for both orientations is given on Fig. 8b. Except reversed phase for certain components of the FMR spectrum the agreement between the calculated and the experimental spectra is fairly good. This confirms validity of the proposed model of anisotropy Eq. (1). The deduced temperature dependence of the magnetic parameters is given in Fig. 9b.

3.5 Temperature Dependence of the Anisotropy Fields

The cumulative data on the magnitudes and temperature variations of the magnetic parameters are given in Fig. 9. All the parameters are given in magnetic induction units (Gauss). It should be recalled that the effective magnetization, M_{eff}, includes perpendicular anisotropy (see Eq. (1)). That is why the values of M_{eff} are essentially reduced compared with the bulk magnetization (~ 1.7 kG [12]). This means that very strong, perpendicular to the film plane, surface anisotropy field (about 5 kG) is induced in the epitaxial ultra-thin film at room temperature. It is also a general feature that both the uniaxial (along the $[1\bar{1}0]$ axis of the GaAs substrate) and the cubic anisotropy fields are significantly large. That is why the anisotropy energy dominates the Zeeman energy and causes such unusual and surprising triple-line FMR spectra in the single ferromagnetic layer. To the authors knowledge, it is the first observation for three FMR lines, all of which correspond to the homogeneous (non-spin-wave) FMR response.

It can be seen from Fig. 9 that all magnetic anisotropy parameters strongly depend on temperature. As the temperature decreases the effective magnetization increases. This is partly due to increase of the saturation magnetization of iron according to the Bloch law, and partly due to decrease of the easy-axis perpendicular anisotropy. The ferromagnetic transition temperature of bulk iron is about 980 K, that is why even at room temperature the magnetic moment is almost fully saturated. Using the literature data on the temperature dependence of iron magnetic moment in the ferromagnetic phase [12] we estimate increase of the iron magnetization in the range from 300 K to 5 K as about 64 G. Obviously, the observed magnitude ($\sim$ 300 G) and almost linear temperature dependence of the effective magnetization do not follow the conventional temperature dependence of saturation magnetization described above. That is why we conclude that essential contribution to the temperature variation of the effective magnetization comes from the temperature dependence of the perpendicular anisotropy. As temperature decreases the perpendicular anisotropy relaxes.

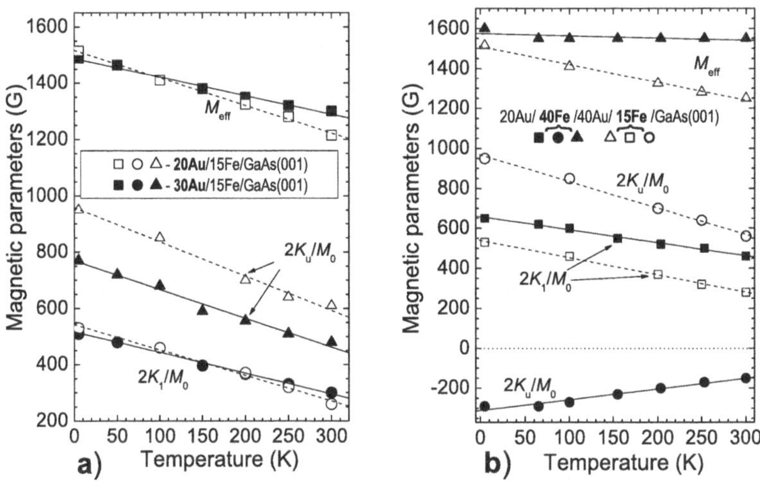

Fig. 9. Temperature dependence of the magnetic parameters: (a) for the single-layer sample; (b) for the double-layer sample.

The absolute values of the both in-plane uniaxial and the cubic anisotropies increase with decreasing the temperature. The sign of cubic anisotropy parameter is positive, making all of the three principal crystalline axes – [100], [010] and [001] – easy for magnetization. However the more strong uniaxial anisotropy, is positive making the $\mathbf{b} = [1\bar{1}0]$ axis a hard direction for magnetization.

As can be noticed from Fig. 9 all of the magnetic anisotropy parameters depend on temperature almost linearly. This implies that there is a common physical reason behind this unified behavior. It is supposed that the strong uniaxial anisotropy in Fe films grown on GaAs is induced by hybridization of the interface electronic states in a ferromagnetic film and the valence electrons in the surface reconstructed GaAs substrate (see, for instance, [3], [13]). Therefore, it is possible that a contraction of interatomic distances, which changes a degree of the metal-substrate hybridization, may also contribute in the temperature dependence of the magnetic anisotropy parameters. However, one can notice from Fig. 9b that for the second iron layer, which is well separated from the substrate by two (first iron and Au) layers, only the perpendicular magnetic anisotropy reveals a change in the temperature behaviour with respect to the first Fe layer. On the other hand, taking into account appreciable differences between the lattice constants for bulk Fe and epitaxially prepared Fe film on the GaAs substrate (see Table 1) one may attribute the temperature dependence of the anisotropy parameters to the linear magneto-elastic effect. It has been proposed that the strain contributes not only to the cubic and perpendicular anisotropies but also to the uniaxial anisotropy due to the anisotropic strain relaxation (see [4], and references therein). In this respect the change in the sign of uniaxial anisotropy for the second iron layer shows that even through the 40ML Au layer the strain-induced anisotropy does not relax to zero. Furthermore, both contributions to the uniaxial in-plane anisotropy mentioned above have to be taken into account to explain the observed behaviour, and the sign of the strain-induced anisotropy is opposite to that of the uniaxial anisotropy due to reconstructed GaAs surface. Therefore, these two contributions counteract each other in the first iron layer. This competition could be relevant for the non-linear behavior of the in-plane uniaxial anisotropy with the iron film thickness reported in the literature [4, 3].

Table 1.

Material (structure)	Thermal expan. $(\times 10^{-6}K^{-1})$			Lattice Parameter	Effect on iron layer
	298 K	23 K	273 K	$a = b = c$ (Å)	
Iron (bcc-Fe)	11.8	5.0	24.0	2.8665	
GaAs(ZnS structure)	5.73			$5.654/2 = 2.827$	compressive strain
Gold (fcc-Au)	14.2	4.6	6.7	$4.078/\sqrt{2} = 2.892$	tensile strain
Chromium (bcc-Cr)	6.2			2.91	large tensile strain
Palladium (fcc-Pd)	11.8	12.2	3.9	$3.891/\sqrt{2} = 2.759$	compressive strain

Actually, there is about -1.5% misfit between lattice parameters of Fe, Au and GaAs substrate. In a result, the Fe film is under a compressive strain. When temperature is lowered down to 4 K from the room temperatures, this stress decreases by about 40 percent, as can be calculated by using the thermal expansion coefficient of Fe, Au and GaAs crystals, given in Table 1. The lattice parameters vary linearly with the temperature, and, in the first approximation, we expect the anisotropy parameters vary linearly with temperature as a result of different thermal expansion coefficients of materials in the contact (see Table 1) and the magneto-elastic coupling. In this case, the cubic anisotropy relaxes to the bulk iron value (about of 530 G) and the strain-induced contribution in the cubic anisotropy is negative in contrast to the bulk one, which is positive and dominating. However, in this picture an increase of absolute value of uniaxial component of anisotropy in the second 40ML iron layer for the 20Au/40Fe/40Au/15Fe/GaAs(001) sample is not clear. Possibly, the step-induced anisotropies in the top of the iron film also contributes in the total value of the uniaxial in-plane anisotropy [3, 14]. A competition between various contributions in the in-plane anisotropies is especially indicative in the case of Cr-cap layer, where the iron layer is under the stretching influence due to larger lattice parameter of Cr compared to Fe. Since the misfit and interface interaction between chromium and iron is much larger compared to the case of contact with noble metals, the strain anisotropies, induced by the stretching influence of Cr cap layer, almost completely cancel the surface-induced uniaxial anisotropy due to the reconstructed GaAs and decrease the absolute value of cubic anisotropy as well.

4 Conclusion

We studied the magnetic anisotropies of epitaxial, ultra-thin iron films grown on the surface reconstructed GaAs substrate. The ferromagnetic resonance technique has been explored extensively to determine magnetic parameters of the films in the temperature range 4–300 K. The triple-peak behavior of FMR spectra was observed for the first time, which allowed accurate extraction of magnetic anisotropies from computer fitting of the model to the experimental data. Presence of the strong perpendicular anisotropy and the uniaxial in-plane anisotropy, induced by the surface reconstruction of the substrate and lattice mismatch with substrate/overlayer, have been deduced from the fitting of the FMR spectra. The switching of the principal anisotropy axes of the top iron layer with respect to the bottom (adjacent to the GaAs substrate) layer has been established in the double-layer sample. It was attributed to the competition of various contributions in the uniaxial anisotropy (surface anisotropy due to the surface reconstruction of the GaAs substrate and strain anisotropy). The surface- and interface-induced origin of the anisotropies has been demonstrated by extensive studies of temperature dependence of anisotropy fields, as well as influence of the cap layer material. The results on temperature dependence have been discussed in terms of the magneto-elastic contribution to the magnetic anisotropies. It was shown that surface reconstruction of GaAs substrate as well as combination of materials in the stack can be used for the tailoring of the magnetic anisotropies in spin-valve-like, double ferromagnetic layer structure.

Acknowledgments

The work was supported by the Gebze Institute of Technology grant No 03-A12-1 and the BRHE grant No REC-007.

References

1. B. Heinrich and J. F. Cochran, Adv. Phys. **42**, 523 (1993).
2. M. Farle, Rep. Prog. Phys. **61**, 755 (1998).
3. T. L. Monchesky, B. Heinrich, R. Urban, K. Myrtle, M. Klaua, and J. Kirshner, Phys. Rev. B **60**, 10242 (1999).
4. S. McPhail, C. M. Gürtler, F. Montaigne, Y. B. Xu, M. Tselepi, and J. A. C. Bland, Phys. Rev. B **67**, 024409 (2003).
5. H. Suhl, Phys. Rev. **97**, 555 (1955).
6. A. G. Gurevich, G. A. Melkov, *Magnetic Oscillations and Waves*, CRC Press, New York, 1996, ch.2.
7. B. Aktaş and M. Özdemir, Physica B **193**, 125 (1994).
8. J. J. Krebs, F.J. Rachford, P. Lubitz, and G. A. Prinz, J. Appl. Phys. **53**, 8058 (1982).
9. Yu. V. Goryunov, N. N. Garifyanov, G. G. Khaliullin, I. A. Garifullin, L. R. Tagirov, F. Schreiber, Th. Mühge, and H. Zabel, Phys. Rev. B **52**, 13450 (1995).
10. Th. Mühge, I. Zoller, K. Westerholt, H. Zabel, N. N. Garifyanov, Yu. V. Goryunov, I. A. Garifullin, G. G. Khaliullin, and L. R. Tagirov, J. Appl. Phys. **81**, 4755 (1997).
11. T. Toliński, K. Lenz, J. Lindner, E. Kosubek, K. Baberschke, D. Spoddig, R. Meckenstock. Solid State Commun. **128**, 385 (2003).
12. *Numerical Data and Functional Relationships in Science and Technology*, Landolt-Börnstein, New Series, vol. III/19A (Springer, Heidelberg, 1986).
13. B. Heinrich, T. Monchesky, R. Urban, J. Magn. Magn. Mater. **236**, 339 (2001).
14. B. Heinrich, Z. Celinsky, J. F. Cochran, A. S. Arrott, K. Myrtle, and S. T. Purcell, Phys. Rev. B **47**, 5077 (1993).

Part IV

GMR Read Heads and Related

New Domain Biasing Techniques for Nanoscale Magneto-Electronic Devices

Z.Q. Lu and G. Pan

School of Computing, Communication and Electronics, University of Plymouth, Plymouth, Devon, PL4 8AA, UK

Summary. A spin valve giant magnetoresistive (GMR) or tunnel magnetoresistive (TMR) head is designed to exhibit a transfer curve, which is sensitive, non-hysteric, linear, and noise free. The basic approach to achieve these properties is to make the easy axis of the free layer parallel to an air bearing surface (ABS) and a longitudinal domain bias technique is employed to magnetise the free layer along its easy axis to saturation so that a single domain state is obtained in the free layer and the head is free from Barkhauson noise during its operation. The most commonly used domain bias techniques are permanent magnet biasing (PM biasing) and antiferromagnet exchange-tab biasing (AF exchange-tab biasing). However, these existing domain bias techniques are becoming obsolete as the continued miniaturization of sensors is approaching the nanometre regime. This paper will first review the basic principles, advantages and disadvantages of the existing domain bias techniques. We will then present our recent work on a new domain bias technique for nanoscale magneto-electronic devices employing interlayer exchange coupling and spin flop of synthetic antiferromagnets.

1 Introduction

A spin valve (SV) or magnetic tunnel junction (MTJ) plays a key role in the present read heads for computer hard disk drives due to their high magnetoresistance (MR) ratio and linear MR response in low fields [1]. A typical SV (MTJ) sensor consists of two ferromagnetic (FM) layers separated by a copper (Al_2O_3, or MgO) layer. One of the FM layers is pinned by an antiferromagnetic (AF) layer along the transverse direction, and the other FM layer, the free layer, is magnetically free. The easy axis of the free layer is in the longitudinal direction parallel to an air bearing surface (ABS) [2, 3]. In spite of its linear response, there exist Barkhauson noises during sensor operation as it lacks a domain control layer. This leads to the development of tail stabilization, in which the read region of the sensor is stabilized in a single domain state by a pair of permanent magnets (PM) or AF exchange-tab on both sides of the read region [1, 4, 5].

In the PM abutted junction domain bias scheme, a pair of permanent magnets, abutted against the ends of sensor, provides a magnetic field that magnetizes the magnetic moment of the free layer along the longitudinal axis to a single domain state [6]. Such a domain bias scheme has been widely used in commercial SV head products. However, the sensitivity of a PM domain bias SV head is limited by the magnetically inactive zones in the vicinity of the PM [7]. As the head size continues to reduce, the inactive zones occupy an increasingly significant portion of the whole sensor, eventually reaching a point at which the head is entirely inactive. The PM domain bias scheme is therefore not suitable for ultra-high density recording heads.

Alternative to the PM domain bias scheme, the free layer of a read head is stabilized by a pair of AF exchange tabs at both ends. In this domain biasing scheme, a longitudinal bias field is provided by a pair of AF exchange tabs which overlap with the two ends of the free layer stripe and generated an exchange bias field in free layer [7, 8]. This domain bias scheme overcomes the inactive zone problem existed in the PM domain bias scheme. However, the striking challenge is that it needs a two-step magnetic field annealing process to achieve the orthogonal orientation of the longitudinal domain bias field and the transverse pinning field. As the orientation of the bias field and the pinning field can only be obtained by magnetic field annealing at above blocking temperature of the AF materials, two AF materials with distinctively different blocking temperature are required in order to minimise the cross interference. The practical difficult is to find the two AF materials which exhibit distinctly different blocking temperatures so that the transverse pinning field orientation firmed during the first step magnetic annealing is not disturbed by the second step magnetic annealing for the formation of the longitudinal domain bias field direction. The blocking temperature of these two AF materials must be sufficiently high to provide an adequate exchange field at the head operating temperature. All these problems make the exchange-tab domain bias scheme difficult to implement in disk drive products.

As the recording density goes further higher, MTJ with MR ratio over 20% has become one of the most promising candidates for a sensor element in hard disk drives [9, 10, 11]. Commercial tunnel magnetoresistive (TMR) heads using Al_2O_3 as a barrier material at 300 Gbits/in^2 has been reported [12]. However, the demand for areal densities over 1 Tbits/in^2 requires an MR ratio significantly higher than 20%.

MTJ with amorphous Al_2O_3 tunnel barrier layers have been extensively studied so far. The MR ratio in such a conventional MTJ can be explained by Julliere's model when experimentally observed spin polarizations are considered [13]. On the other hand, ab initio calculations have predicted extremely large MR ratios in Fe (001) /MgO (001)/Fe (001) single-crystalline MTJs as the result of coherent tunneling [14, 15]. Experimental studies performed on epitaxial Fe (001) /MgO (001)/Fe (001) have shown that their MR ratios have reached as high as 188% at room temperature far beyond those obtained with conventional MTJs [16, 17]. These MTJs have grown on single-crystal substrates, which limits their practical application. An MR ratio of over 200% at RT has been achieved in MTJs with a structure of FeCo(B) /MgO (001)/

CoFe(B) [18, 19]. However, the RA of the MTJs was about 420 $\Omega\mu m^2$, which is too high for TMR head application. Recently published results using MgO as a barrier material has shown TMR ratios in excess of 100% with an RA of 2 $\Omega\mu m^2$, providing more opportunity for extending TMR recording head applications [20].

However, the current perpendicular to plane (CPP) geometry in MTJs gives rise to an electrical short in the case of the conventional tail stabilization. Several longitudinal domain bias schemes have been suggested to avoid the electrical short between a free layer and pinned layer [21, 22, 23, 37], in which the AF bias layer is placed on the top of the free layer rather than at the ends of the reader active area. However, they still have not overcome the problem that needs two AF materials with distinctly different blocking temperatures to achieve the orthogonal orientation of the longitudinal domain bias field and the transverse pinning field.

Synthetic spin-filter spin valve with spin-engineered bias scheme "substrate /Ta/NiFe/IrMn/NiFe/NOL/Cu_1/CoFe/Cu_2/CoFe/Ru/CoFe/IrMn/Ta" was developed from our recent work on spin flop in synthetic spin valve and interlayer exchange domain bias schemes [23–29], [37]. In this structure, the orthogonal magnetic configuration for biasing and pinning field is obtained by a one-step magnetic annealing process by means of spin flop, which eliminates the need for two AF materials with distinctly different blocking temperatures. The longitudinal domain stabilization of the free layer can be achieved by interlayer coupling field through Cu_1 spacer. By adjusting the thickness of Cu_1 layer, the interlayer coupling bias field can provide domain stabilization and is sufficiently strong to constrain the magnetization in coherent rotation. Such a domain bias scheme will have important technological application in nanoscale SV and MTJ read heads.

2 Review of existing domain biasing techniques – basic principles and limitations

2.1 Spin valve sensor with PM abutted junction domain bias scheme

A typical spin valve sensor consists of a ferromagnetic free layer and a ferromagnetic pinned layer separated by a thin spacer, as shown in Fig. 1 [2, 3]. The magnetic moment of the pinned layer is generally fixed along the transverse direction by exchange coupling with an AF layer such as FeMn, NiMn, IrMn or PtMn, while the magnetic moment of the free layer is allowed to rotate in response to signal fields. The spin valve response is given by

$$\Delta R \ \propto \ \cos(\theta_1 - \theta_2) \propto \ \sin\theta_1 \tag{1}$$

where θ_1 and $\theta_2 (= \pi/2)$ represent the direction of the free and pinned layer magnetic moments respectively (Fig. 1). If the uniaxial anisotropy hard axis of the free layer is oriented along the transverse direction, then the magnetic signal response is linear ($\Delta R \propto H$). However, there exist Barkhauson noises during sensor operation as small geometry MR sensors exhibit spontaneous tendency to break up into complicated multi-domain states. Among many factors, the shape demagnetization effect is the primary cause for multi-domain formation. This understanding leads to the development of tail stabilization, in which the read region of the sensor is stabilized in a single domain state by a pair of permanent magnets or AF exchange-tab on both sides of the read region [1, 4, 5].

In the PM abutted junction domain bias scheme shown in Fig. 1, a pair of permanent magnets (Cr/CoPtCr), abutted against the ends of a sensor, provide a magnetic field that magnetizes the magnetic moment of the free layer along the ABS to a single domain state [6]. In order for this scheme to succeed, the coercivity (H_c) of the PM layer must be high enough to pin the free layer in the tail and the remanence magnetic moment ($M_r t$) large enough to saturate the magnetic moment of the free layer along the longitudinal axis. It is crucial to control its magnetic moment for optimizing sensor stability and sensitivity behaviours. The $M_r t$ of the PM layer in the tail region is preferably higher than that of the free layer in the read region to ensure high read sensitivity. The ratio of these moments is typically between 2 and 5. This mismatch of the magnetic results in stray magnetic flux generated at each side of the reader by the permanent magnet.

Despite the difficulties in controlling the magnetic properties of PM film and complicated at abutted junctions, the simplicity of the sensor fabrication makes this scheme attractive. The PM abutted junction domain bias scheme has been widely used in commercial SV head products. However, the sensitivity of a PM domain bias SV head is limited by the magnetically inactive zones in the vicinity of PM [7]. As the head size continues to reduce, the inactive zones occupy an increasingly significant portion of the whole sensor. The PM domain bias scheme will therefore not suitable for ultra-high density recording heads.

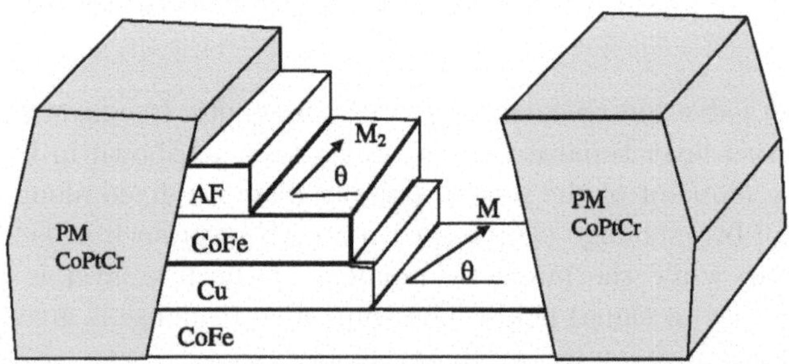

Fig. 1. Schematic of spin valve sensor with PM abutted junction domain bias scheme

2.2 Spin valve sensor with exchange tab domain bias scheme

An alternative to the PM abutted junction domain bias scheme is to use exchange biasing, called AF exchange tab domain bias scheme. In this domain bias scheme, a pair of AF exchange tabs (such as FeMn, NiMn, PtMn, IrMn, et al), is deposited on top of the free layer at both sides, as shown in Fig. 2. As a result, a fairly large effective bias field is created through interfacial exchange coupling, and this bias field can be aligned longitudinally along the sensor for domain suppression. In this case, the width of the inactive region is determined by the strength of the direct exchange coupling in the free layer, which is short range. Thus, there is no inactive zone problem in this domain bias scheme. The magnetic moment (M_st) of the free layer tends to decrease with increasing areal density, resulting in higher bias fields. In addition, the absence of the abutted junction removes a potential source of stress and parasite resistance, improved the signal to noise ratio. There is also no flux mismatch at the edge of the sensor as the free layer is continuous across the edge of the reader. Spin valve sensors with AF exchange tab domain bias scheme exhibit much higher read sensitivity than that with PM domain bias scheme at deep submicron region. There are difficult issues yet to be solved. A potential drawback is that the free layer in the wing area picks up flux from adjacent tracks and adds significant side reading. The real challenge is that there needs two-step magnetic field annealing process to achieve the orthogonal orientation of the longitudinal domain bias field and the transverse pinning field. As the orientation of the bias field and the pinning field can only be obtained by magnetic field annealing at above blocking temperature of the AF materials, two AF materials with distinctively different blocking temperature are required in order to minimise the cross interference. The practical difficult is to find the two AF materials which exhibit distinctly different blocking temperatures so that the transverse pinning field orientation is not disturbed by the second step magnetic annealing for the formation of the longitudinal domain bias field direction. The blocking temperature of these two AF materials must be sufficiently high to provide an adequate exchange field at the head operating temperature. All these problems made the exchange-tab domain bias scheme difficult to implement in disk drive products.

2.3 Spin valve sensor with interlayer exchange coupling domain bias scheme

MTJs with MR ratio over 40% have recently become one of the most promising candidates for a sensor element in hard disk drives [18, 19, 20] . The conventional tail stabilization scheme is not suitable for TMR heads as the CPP geometry in MTJ gives rise to an electrical short. Several longitudinal domain bias schemes have been suggested to avoid the electrical short between a free layer and pinned layer [21, 22, 23, 27], in which the AF bias layer is

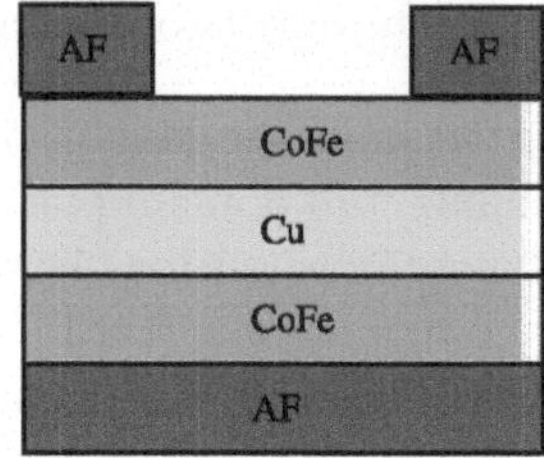

Fig. 2. Schematic of spin valve sensor with AF exchange tab domain bias scheme

placed on the top of the free layer rather than at the ends of the active sensor area. One of the domain bias schemes is the interlayer exchange coupling domain bias scheme with structure of sub/Ta/NiFe/FeMn/NiFe (BL)/Mo/NiFe (FL)/Cu/NiFe (PL)/CrMnPt/Ta. The magnetic domains of the free layer can be stabilized by using the interlayer coupling from the NiFe (BL) biasing layer through the Mo layer. We choose Mo as the interlayer spacer for the bias scheme because NiFe/Mo multilayers exhibit a modest interlayer coupling without giant magnetoresistance effect [38]. The top AF (CrMnPt) layer has a blocking temperature (T_b) of $\sim$ 320°C, far higher than that of the bottom AF (FeMn) layer ($T_b \sim$ 140°C). These features allow for a two-step annealing with magnetic filed in orthogonal directions. Planar Hall effect (PHE) was used to study the free layer reversal mechanism.

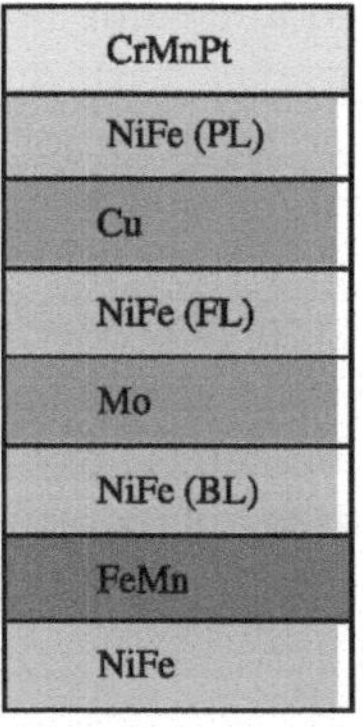

Fig. 3. Schematic of spin valve sensor with interlayer exchange coupling domain bias scheme

Spin valves with conventional structure of sub/Ta 4 nm/ NiFe 7 nm/ Cu 3 nm/ NiFe 3.5 nm/CrMnPt 30 nm/ Ta 5 nm (sample 1) and with new structure of sub/Ta 4 nm/ NiFe 2 nm/FeMn 6 nm/Mo tMo /NiFe 7 nm/ Cu 3 nm/ NiFe 3.5 nm/CrMnPt 30 nm/ Ta 5 nm (sample 2) were deposited in a magnetron sputtering system with a base pressure of 2.0×10^{-5} Pa. t_{Mo} was optimized for optimum bias field. Sample annealing was carried out in a vacuum furnace at a base pressure of 2.0×10^{-4} Pa. The magnetic field for annealing was 500 Oe. Sample 1 was annealed at 320°C for 1 h in a transverse direction (perpendicular to ABS direction). Sample 2 was magnetically annealed in two steps, initially at 320°C for 1 h in a transverse direction, then cooled down to 140°C, further annealed at 140°C for 0.5 h in a longitudinal direction, and then cooled to room temperature.

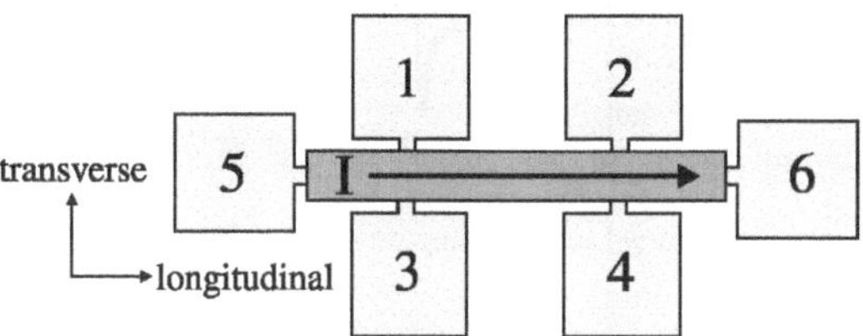

Fig. 4. Schematic illustration of the electrical connections for measurement of PHE and MR in a patterned spin valve element.

The samples for MR and PHE measurements were patterned into six terminal bars as shown in Fig. 4. The distance was 200 mm between the PHE terminals 1 and 3 (or 2 and 4), and 1 mm between the MR terminals 3 and 4 (or 1 and 2). The easy axis of FL was aligned to the longitudinal direction. MR(H) and PHE(H) of each sample were measured simultaneously at room temperature in a sweeping field of 100 Oe applied in the film plane at different angle (a) with respect to the easy axis of the FL. The sensing current for MR(H) and PHE(H) measurement was 1 mA.

The MR(H) responses of sample 1 at filed angles of 75°, 90° and 105° were shown in Fig. 5(a), 5(d) and 5(g), respectively. The MR(H) curves showed a typical linear response and their shapes were not sensitive to α. However, the PHE(H) curves, shown in Fig. 5(b), 5(e) and 5(h), were sensitive to α. A small change in a led to a different PH(H) response. The MR(H) curves of sample 2 at field angles of 75°, 90°, and 105° were shown in Fig. 6(a), 6(d), and 6(g), respectively, which showed a typical linear response as sample 1. The PHE(H) of sample 2 , shown in Fig. 6(b), 6(e) and 6(h) were not sensitive to a, which is in contrast to sample 1.

To understand the observed the MR and PHE response, a model based on Boltzmann transport equation was used to characterize the MR and PHE voltages in spin valves. The conductivity tensors σ_{ij} (the diagonal part determines the MR voltage, and the off-diagonal part determines PHE voltage) are given by

$$\sigma_{ij}^{\uparrow(\downarrow)} = \frac{ne^3}{2mv_f}\frac{3}{4\pi t}\int dz \int d^3\widehat{v}\ \widehat{v}_i\widehat{v}_j\ \lambda_{eff}^{\uparrow(\downarrow)}(\overleftarrow{v}, z) \tag{2}$$

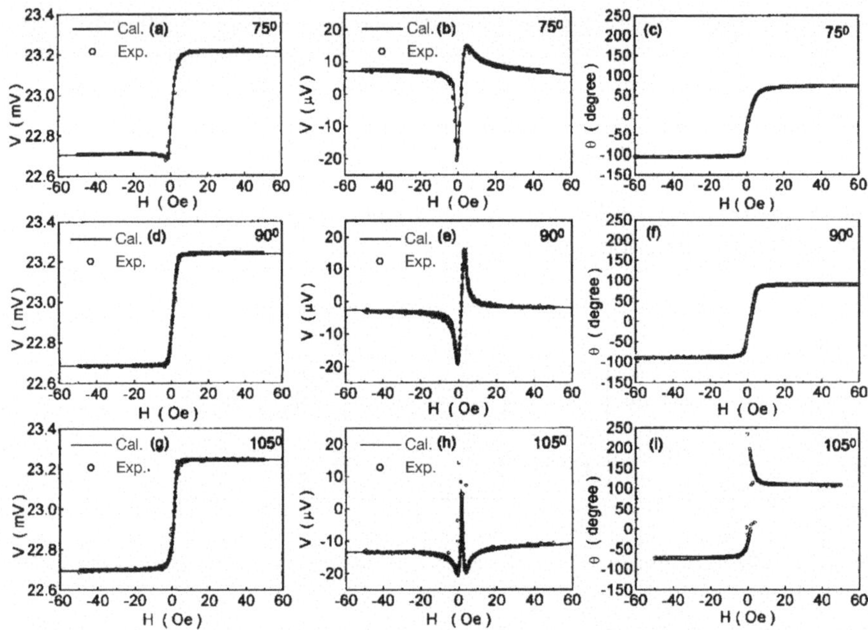

Fig. 5. Experimental (dot) and calculated (line) MR and PHE responses of Sample 1 in applied magnetic field at different a. First column shows the MR curves. The second column the PHE curves and the third column the corresponding simulated magnetization angles of the free layer

where n is the total conduction electron density, e and m are the electron charge and mass, respectively, and v_f is Fermi velocity. $\lambda_{eff}^{\uparrow(\downarrow)}$ is the effective electron mean free path. Here we assume a dependence of the mean free path on the angle θ between the electron velocity $\overrightarrow{v}$ and the magnetization $\overleftarrow{M}$,

$$\lambda_{eff}^{\uparrow(\downarrow)} = \lambda_0^{\uparrow(\downarrow)}(1 - a^{\uparrow(\downarrow)} \cos^2 \theta - b^{\uparrow(\downarrow)} \cos^4 \theta) \tag{3}$$

where parameter $a^{\uparrow(\downarrow)}$ and $b^{\uparrow(\downarrow)}$ are a measure for the anisotropy of the scattering, which are determined by the amplitude of MR voltage. Supposing that the magnetization process of FM layers could be treated by the Stone-Wolfarth model, then the magnetic energy per unit surface area can be written as follow:

$$\begin{aligned} E = K_{ub}t_b \sin^2 \theta_b - M_s t_b H_b \cos(\varphi_b - \vartheta_b) - M_s t_b H \cos(\alpha - \vartheta_b) \\ + K_{uf} t_f \sin^2 \theta_f - M_s t_f H \cos(\alpha - \vartheta_f) \\ + K_{up} t_p \sin^2 \theta_p - M_s t_p H_p \cos(\varphi_p - \vartheta_p) - M_s t_p H \cos(\alpha - \vartheta_p) \\ + J_1 \cos(\theta_f - \theta_p) - J_2 \cos(\theta_f - \theta_b) \end{aligned} \tag{4}$$

Here θ_f, θ_p, and θ_b are the angles of the FL, PL and BL magnetization with respect to the easy axis direction of free layer, respectively. t_f, t_p and

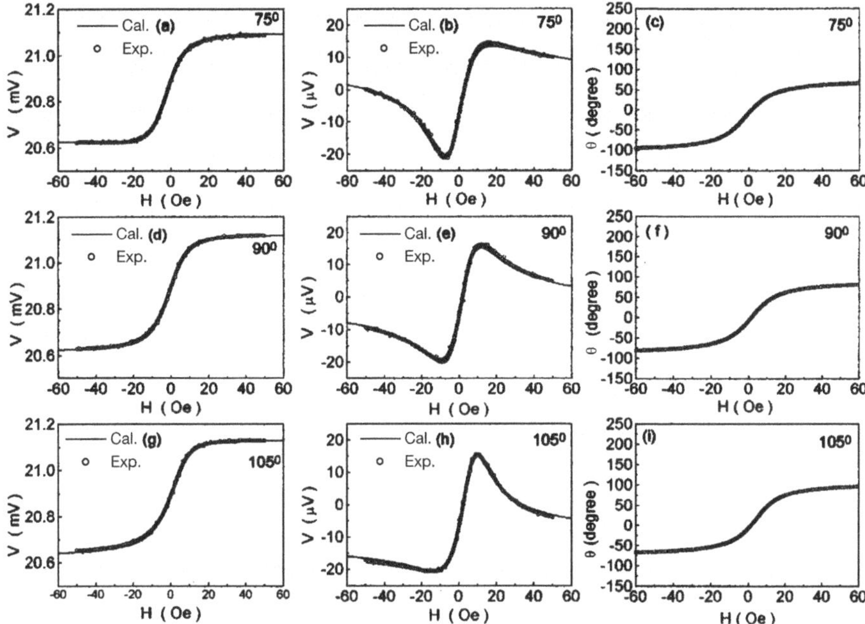

Fig. 6. Experimental (dot) and calculated (line) MR and PHE responses of Sample 2 in applied magnetic field at different α. First column shows the MR curves. The second column the PHE curves and the third column the corresponding simulated magnetization angles of the free layer

t_b the thickness, and K_{uf}, K_{up} and K_{ub} the effective anisotropy constant. J_1 is the interlayer coupling between FL and PL layers, J_2 is the interlayer coupling between FL and BL layer through thin Mo layer. M_s is the saturation magnetization of the FM layers. H_p and H_b are effective pinning fields of the PL and BL, respectively.

In the numerical calculation, we use the value $\lambda^{\uparrow}_{eff} = 4.4$ nm, $\lambda^{\downarrow}_{eff} = 0.6$ nm, $a^{\uparrow} = 0.0327$, $a^{\downarrow} = -0.00556$, $b^{\uparrow} = b^{\downarrow} = 0$, $\rho_{NiFe} = 15$ μΩ cm and $M_s = 800$ emu/cm^3 for NiFe. The calculated MR(H) and PHE(H) fitted the experimental results.

For sample 1, $\varphi_p = -88°$ was obtained by curve fitting. The magnetization orientations of FL as a function of applied field, represented by the $\theta_f(H)$ curve, were obtained from the numerical calculations and were shown in Figs. 5(c), 5(f) and 5(i) for $\alpha = 75°$, $90°$ and $105°$, respectively. Two distinct reversal modes were obtained from the simulation results. Figures 5(c) and 5(f) showed the smooth changes of θ_f with the applied field, which suggested that the magnetization of the FL rotated coherently with applied field. This type of free layer reversal is labeled type-C reversal [36], corresponding to the coherent magnetization rotation, which was Barkhausen noise free. The

$\theta_f(H)$ curves in Fig. 5(i) showed a discontinuous jump of θ_f from 195° to 225° at the applied field around 1.0 Oe, suggesting that there existed abrupt domain wall motion during the magnetization reversal process. This type of free layer reversal is labeled type-A reversal [27]. The magnetization jump gave rise to Barkhausen noise in a spin valve sensor despite the linear MR response of the sensor. For sample 2, $\varphi_p = -87°$ and $\varphi_b = -2°$ were obtained by curve fitting. The Magnetization orientation of FL as a function of applied field, represented by $\theta_f(H)$ curve, was shown in figs. 6(c), 6(f), and 6(i) for a of 75°, 90° and 105°, respectively. Only the reversal mode of type-C was observed for this sample for $60° < \alpha < 120°$. The free layer appeared to remain in the single domain state in all three cases. It is evident that the interlayer coupling bias field from the BL effectively stabilized the magnetic domain of the FL of the spin valve sensor. The magnetization reversal of the FL was free from Barkhausen jumps. The major advantages of such a bias scheme are that there is no "dead zone problem" as in PM biasing, and the strength of bias field can be controlled by varying the thickness of the Mo spacer.

2.4 Spin valve sensor with long-range exchange coupling domain bias scheme

Alternative overlaid domain bias scheme was developed from the long-range exchange coupling [21, 22, 37], shown in Fig. 7. Exchange coupling is first observed in ferromagnets in contact with antiferromagnets [30]. It is believed that exchange bias originates from the local exchange interaction between the FM moments and the AF moments across the interface. Recently, Long-range exchange coupling between a ferromagnet and an antiferromagnet across a nonmagnetic layer was observed in a trilayer of NiFe/Cu, Ag, and Au/NiO [31, 32]. Figure 8 shows the coercivity (H_c) of the CoFe layer and the exchange-coupling field (H_e) between the CoFe and IrMn across the Cu layer for various thickness (t_{Cu}). The exchange-coupling field is found to decrease monotonically with increasing t_{Cu}. These results imply that the bias field of a free layer could be easily adapted for sensitivity control and stability improvement by altering t_{Cu}. This domain bias scheme is promising in the application of STJ and CPP-SV. However, there still faces the problem that two AF materials with distinctly blocking temperature are needed to achieve the orthogonal orientation of the longitudinal domain bias field and the transverse pinning field.

3 Challenges of domain biasing for nanoscale devices

In order for SV and STJ sensors to perform magnetic recording at ultrahigh areal densities beyond 100 Gb/in^2, their thickness, width and height have been progressively reduced to nanoscale. The minaturization of the sensors causes substantial increases in damagnetization fields at sensor edges and

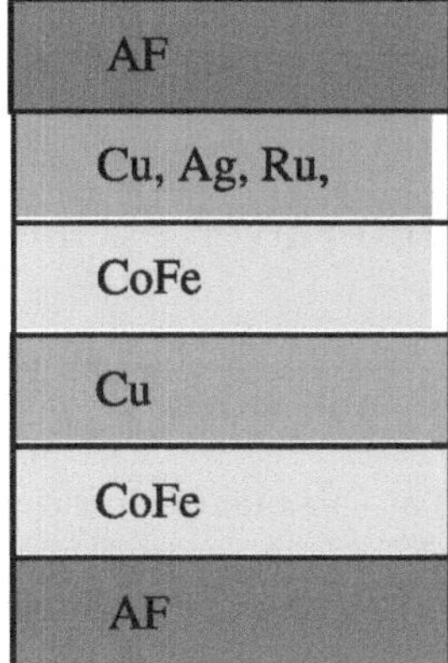

Fig. 7. Schematic of spin valve sensor with long-range exchange coupling domain bias scheme

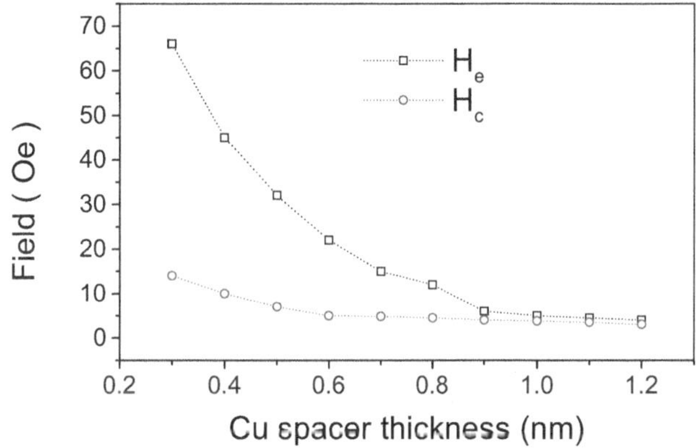

Fig. 8. Cu spacer thickness dependence of the exchange field and the coercivity in CoFe/Cu/IrMn trilayer

move severe multidomain activities in the sense layers. It thus becomes more challenge to stabilize the sense layers. The new domain bias schemes need to satisfy the following conditions:

a) Simple annealing process to achieve magnetic orthogonal configuration between bias field and pinning field. The domain bias field for sense layer parallel to ABS direction and the pinning field for pinned layer perpendicular to ABS direction;
b) No magnetic inactive zones in the sense layer;
c) Well defined trackwidth, no side reading problem;
d) Bias field strength easy to control;

e) No electrical shorting problem for STJ sensors;
f) Less complicated film deposition and microfabrication process for heads.

4 New techniques – interlayer exchange coupling and spin flop biasing

4.1 Spin flop in synthetic spin valve

In simple spin valve sensors at sub-micrometer scale, the magnetostatic field arising from the pinned layer leads the magnetization of the free layer canted away from the longitudinal direction. The canted magnetization in the free layer yields amplitude asymmetry and limits sensor dynamic range [24]. In order to eliminate the interlayer magnetostatic fields, a synthetic antiferromagnet (SAF) is adopted [34, 35]. Synthetic spin valves consist of FL/Cu/P1/Ru/P2/AF, where FL is the free layer, P1 and P2 are the two FM layers antiferromagnetically coupled through an ultrathin Ru layer owing to an RKKY interlayer exchange coupling. P2 is exchange biased by an adjacent AF layer shown in Fig. 9. In this structure, the antiparallel alignment of the magnetization in P1 and P2 reduces the magnetostatic field created by the pinned layers on the free layer. Another advantage is the large effective exchange pinning field even at high temperature, which results in magnetic stability of the sensors. An interesting spin-flop phenomenon was also theoretically predicted on these spin valves [24] and experimentally demonstrated [25].

The SAF structure in synthetic spin valves, e.g. CoFe (I)/Ru/CoFe (II)/PtMn, employs the interlayer AF coupling in the CoFe (I)/Ru/CoFe (II) tri-layer structure to obtain a high pinning field, and the interfacial exchange coupling in CoFe (II)/PtMn films to provide a fixed spin orientation. The RKKY interlayer coupling in CoFe (I)/Ru/CoFe (II) tri-layer varies in strength as a function of the Ru thickness ranging from 0.3 nm to 1 nm and the thickness of the two CoFe layers

The interfacial exchange coupling and its spin orientation in CoFe(II)/PtMn bi-layer can be introduced at above blocking temperature. Consider an SAF with two identical ferromagnetic layers under a uniform field, shown in Fig. 10. The axis along which the magnetizations of the two layers are antiparallel to each other is referred to as the antiparallel (AP) axis and ϑ is the angle between the AP axis and the external field direction. Denoting as the deviation angle of each layer's magnetization away from AP axis, the surface energy area density can be written as:

$$E = -A\cos(2\alpha) - 2M_s H\delta\sin\theta\sin\alpha \tag{5}$$

where A is the surface interlayer antiferromagnetic exchange constant, M_s the saturation moment and δ the thickness of each layer. Assuming $\alpha \ll 1$

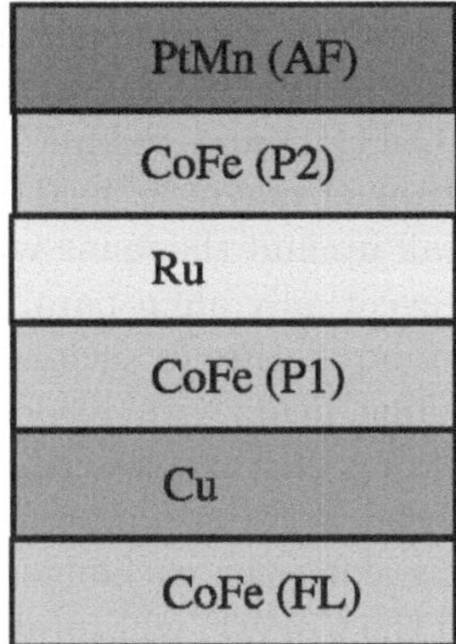

Fig. 9. Schematic of spin valve with synthetic antiferromagnet

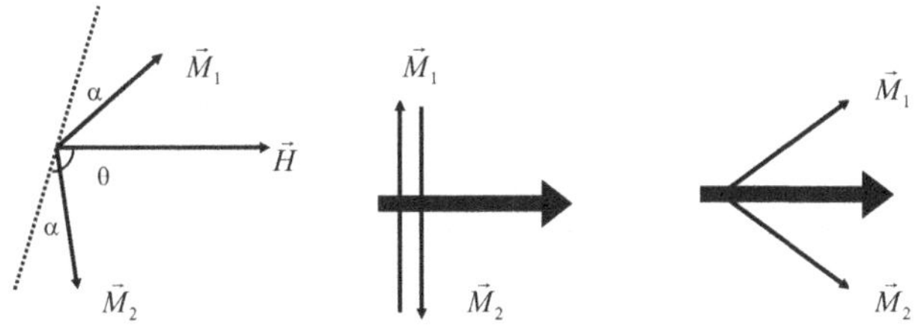

Fig. 10. The field on a SAF structure

when interlayer exchange-coupling field is far larger than the applied external field, the energy density can be written as:

$$E = -A + 2A\alpha^2 - 2M_s H\delta \sin\theta \bullet \alpha \tag{6}$$

From above equation, it is seen that the Zeeman energy of the system can be reduced with the magnetization of each P layer slightly deviating from AP axis towards the field direction. This energy reduction becomes maximum when the AP axis orients orthogonal to the field. Using equilibrium condition to solve α, we obtain:

$$E = -A - \frac{(M_s H\delta)^2}{2A} \sin^2\theta \tag{7}$$

The reduction of the Zeeman energy over the exchange energy manifests itself as an effective uniaxial anisotropy with easy axis orthogonal to the field direction in film plane. This is similar to the spin flop phenomenon in an antiferromagnetic material. Therefore, for a SAF structure with two identical ferromagnetic layers (CoFe (I)/Ru/CoFe (II)/PtMn), the SAF is unpinned before annealing. When this SAF is annealed to above 270°C under a small magnetic field, the AP axis of the unpinned SAF will rotate to a direction perpendicular to the axis of the magnetic field. If sample cools down from

270°C after annealing 2 hours under same magnetic field, the spin orientation of CoFe (II) at 270°C is preserved to a lower temperature because of the exchange coupling between CoFe(II) and PtMn, which can give direct evidence of spin flop. When a stronger magnetic field is applied, the antiparallel coupling in the SAF will break up and the spins will scissor toward the axis of the applied field and will eventually align parallel to the field axis. If the SAF is cooled down to room temperature in such a condition, either a canted pinning field or parallel pinning field, with respect to the axis of applied field, can be obtained depending on the strength of the applied annealing field. When the applied removed at room temperature, the spin of CoFe(II) is pinned by the pinning field along the exchange pinning direction formed in the annealing process, and the CoFe(I) will rotate back and in antiparallel orientation with CoFe(II) due to the very strong antiferromagnetic coupling in the CoFe(I)/Ru/CoFe(II). Canted pinning field has been experimentally observed.

4.2 Effect of Ru thickness on spin flop in synthetic spin valves

The dependence of exchange pinning angle and magneto-transport properties of synthetic spin valves with structure of sub/Ta/NiFe/IrMn/P2/ Ru/ P1 /Cu / CoFe /Ta on Ru thickness have been systematically studied both experimentally and theoretically in our recent work [29]. Samples were deposited by a PVD system and annealed in a magnetic field of 6 kOe parallel to the easy axis of the free layer.

Assuming that the reversal of spin valves can be treated by Stoner-Wohlfarth model, the magnetic energy per unit area can be written as follows:

$$\begin{aligned} E = {} & K_{uf} t_f \sin^2\theta_f - M_f H \cos(\alpha - \theta_f) - J_1 \cos(\theta_f - \vartheta_{p1}) \\ & + K_{up1} t_{p1} \sin^2\theta_{p1} - M_{p1} H \cos(\alpha - \theta_{p1}) - J_2 \cos(\theta_{p1} - \theta_{p2}) \\ & + K_{up2} t_{p2} \sin^2\theta_{p2} - M_{p2} H \cos(\alpha - \theta_{p2}) - J_3 \cos(\beta - \theta_{p2}) \end{aligned} \tag{8}$$

Here θ_f, θ_{p1}, and θ_{p2} are the angles of the FL, P1 and P2 magnetization with respect to the easy axis direction of free layer, respectively. t_f, t_{p1} and t_{p2} the thickness, M_f, M_{p1} and M_{p2} the magnetic moment, and K_{uf}, K_{up1} and K_{up2} the effective anisotropy constant. J_1 is the interlayer coupling between FL and P1 layers, J_2 is the interlayer coupling between P1 and P2 layer through thin Ru layer. J_3 is the exchange biasing energy. β is the angle between the exchange bias field and easy axis of the free layer. The shapes of measured MR curves and effective exchange field H_{pin} depend strongly on the thickness of Ru. Typical experimental (dots) and simulated (line) MR curves for spin valves with SAF structure of Ta 3.5 nm/ NiFe 2 nm/ IrMn 10 nm/CoFe 4 nm/Ru tRu/CoFe 4 nm/ Cu 2.5 nm/ CoFe 0.8 nm/ NiFe 9 nm/ Ta 10 nm are shown in Figs. 11(a), 11(d) and 11(g) for Ru thickness of 0.4 nm, 0.8 nm and 1 nm, respectively, where interlayer coupling changes from strong AF coupling to ferromagnetic coupling. During the MR measurement, the field was applied along the free layer easy axis.

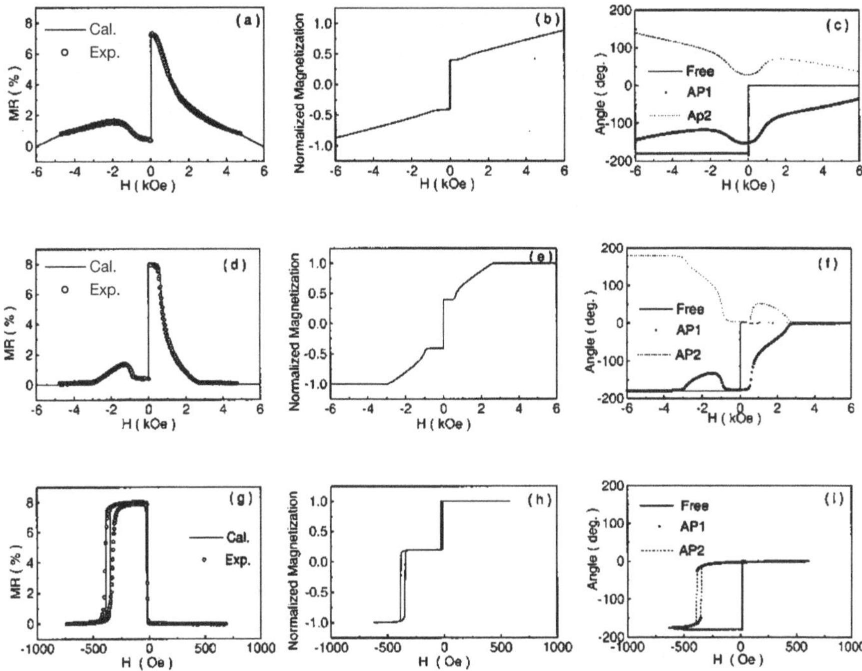

Fig. 11. MR(H) and M(H) curves and the corresponding magnetization angles of synthetic spin valves with different Ru thickness. MR(H) curves for $t_{Ru} = 0.4$, 0.8 and 1 nm are shown in (a), (d) and (g), respectively; M(H) curves for t_{Ru}=0.4, 0.8 and 1 nm in (b), (e) and (h). The corresponding magnetization angles for the free and pinned layers for $t_{Ru} = 0.4$, 0.8 and 1 nm in (c), (f) and (i).

In the numerical calculation, we use the value $\lambda_0^{\uparrow} = 10.5$ nm, $\lambda_0^{\downarrow} = 0.5$ nm, $a^{\uparrow} = 0.0327$, $a^{\downarrow} = -0.00556$, $b^{\uparrow} = b^{\downarrow} = 0$, and $M_s = 1543$ emu/cm^3 for CoFe, value $\lambda_0^{\uparrow} = 5$ nm, $\lambda_0^{\downarrow} = 0.58$ nm, and $M_s = 800$ emu/cm^3 for NiFe. The values of the fitting parameters $K_{uf} = 8 \times 10^3$ erg/cm^3, $K_{up1} = 8 \times 10^3$ erg/cm^3, $K_{up2} = 2.4 \times 10^4$ erg/cm^3, $J_1 = 1.2 \times 10^{-3}$ erg/cm^2, and $J_3 = 0.128$ erg/cm^2, are chosen as independent of the Ru thickness, while J_2 depends on t_{Ru}. The simulated MR responses are shown in Figs. 11(a), 11(d) and 11(g), and the magnetization curves in Figs. 11(b), 11(e) and 11(h). As shown in the figures, the model is able to reproduce the experimental observations to a remarkable degree of accuracy. The magnetization reversal can be understood in more detail from Fig. 11(c), 11(f), and 11(i), which show the variation of the angles (θ_f, θ_{p1}, θ_{p2}) as a function of the applied fields.

Figure 12 shows the interlayer exchange coupling J_2 and pinning angle as a function of Ru thickness. As shown in the figure, the SAF layer have the strongest AF coupling at $t_{Ru} = 0.3$ nm, $J_2 = 1.7$ erg/cm^2. It is noted that the pinning direction is not along the easy axis of the free layer, the pinning angle β is about 70° with respect to the easy axis of the free layer. The misalignment of the pinning direction from the easy axis of the free layer was caused by the spin flop during the magnetic annealing process. Because of the very strong interlayer AF coupling, the magnetic field of 6 kOe applied during annealing was not strong enough to overcome the torque produced by the interlayer AF coupling to align the P1 and P2 along the easy axis of the free layer. The pinning angle is determined by the vector sum of the interlayer antiferromagnetic coupling field and the external field during annealing.

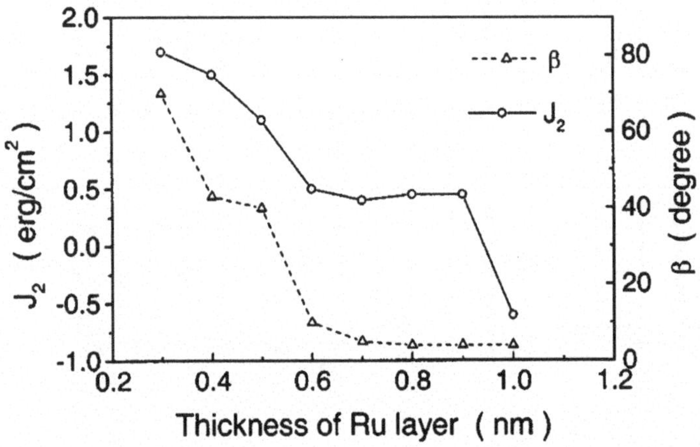

Fig. 12. Interlayer exchange coupling J_2 and the pinning angle β as a function of Ru thickness for samples used in this work annealed under 6 kOe

4.3 Spin valves with spin-engineered domain-biasing scheme

A synthetic spin-filter spin valve with spin-engineered domain bias scheme was developed from recent work on spin flop in synthetic spin valve and interlayer exchange domain bias [29]. A typical layer structure is Ta/B1 /AF1 /B2 /NOL/Cu1 /FL/Cu2 /P1 /Ru/P2 /AF2 /Ta, as schematically shown in Fig. 13, where B1 and B2 are NiFe ferromagnetic layers, P1 and P2 are CoFe pinned ferromagnetic layers, AF1 and AF2 are antiferromagnetic layers, typically PtMn, IrMn, or NiMn. The magnetic domains of the FL can be stabilized by the interlayer coupling field from the B2 biasing layer through the Cu_1 which functions as a spin filter [26] as well as interlayer coupling layer. NOL is a nano-oxide specular reflection layer used to enhance the MR ratio of the spin valve and to prevent the giant MR (GMR) effect at the Cu_1 interface. In this scheme, the two antiferromagnetic layers for biasing and pinning can be of the same material. This eliminates the requirement for two AF materials with distinctly different blocking temperatures. By one-step magnetic annealing process, we can realize a bias field along the annealing field and the pinning field orthogonal to the annealing field by means of a spin-flop effect.

Synthetic spin-filter spin valve (Sample 3) with structure of sub/ Ta 3 nm/NOL /Cu 1.2 nm/CoFe 2 nm/Cu 2.5 nm/CoFe 2.5 nm/Ru 0.5 nm/CoFe 2.5 nm/IrMn 6 nm/Ta 3 nm and synthetic spin-filter spin valve with spin-engineered domain bias scheme (sample 4) with structure of sub/Ta 3 nm/ NiFe 2 nm/IrMn 4 nm/NiFe 2 nm/NOL/Cu1 1.2 nm/CoFe 2 nm/Cu_2 2.5 nm/ CoFe 2.5 nm/Ru 0.5 nm/CoFe 2.5 nm/IrMn 8 nm/Ta 3 nm were prepared in a magnetron sputtering system with a base pressure of 2.03×10^{-6} Pa. The thickness of the Cu1 spacer was optimized for optimum biasing field and MR ratio. Films were deposited in a uniform magnetic field of 100 Oe.

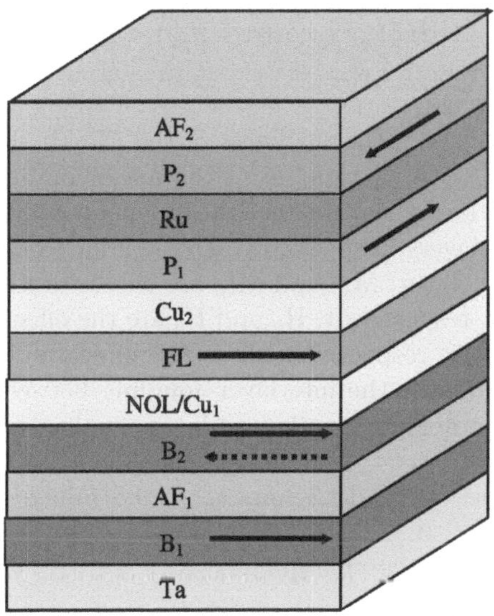

Fig. 13. Schematic diagram of the layer structure and magnetic configuration of synthetic spin-filter spin valve with spin-engineered domain bias scheme

Sample annealing was carried out in a furnace in Ar ambient. The magnetic field for annealing was of 1 kOe. Samples were annealed at 270°C for 30 min in a longitudinal field. A synthetic antiferromagnet (CoFe 2 nm/Ru 05 nm/CoFe 2 nm/ IrMn 8 nm) was used in both samples, the P_1 and P_2 layers were chosen to the same thickness to establish the pinning field perpendicular to applied magnetic annealing field at low field by means of spin flop.

The MR(H) responses of sample 3 at field angle α of 80°, 90°, and 100° were shown in Figs. 14(a), 14(d), and 14(g), respectively. The MR(H) curves showed a typical linear response and were not sensitive to field angle α, and the PHE(H) curves in Figs. 14(b), 14(e), and 14(h) were sensitive to α. A small change in a led to a different PHE(H) response. The MR(H) responses of sample 4 at field angles of 80°, 90°, and 100° in Figs. 15(a), 15(d), and 15(g), respectively also showed a typical linear response as same as sample 3. However, the shapes of the PHE(H) curves of sample 4 in Figs. 15 (b), 15(e), and 15(h) were not sensitive to a, which is in contrast to sample 3.

Supposing that the magnetization process of FM layers is treated by Stoner-Wohlfarth model, then, the magnetic energy per unit surface area can be written as follows:

$$\begin{aligned} E = K_{ub2}t_{b2}\sin^2\vartheta_{b2} - HM_{sb}t_{b2}\cos(\alpha-\vartheta_{b2}) - H_bM_{sb}t_{b2}\cos(\varphi_b-\vartheta_{b2}) \\ +K_{uf}t_f\sin^2\vartheta_f - HM_st_f\cos(\alpha-\vartheta_f) \\ +K_{up1}t_{p1}\sin^2\vartheta_{p1} - HM_{sp}t_{p1}\cos(\alpha-\vartheta_{p1}) \\ +K_{up2}t_{p2}\sin^2\vartheta_{p2} - HM_{sp}t_{p2}\cos(\alpha-\vartheta_{p2}) - H_pM_{sp}\cos(\varphi_p-\vartheta_{p2}) \\ -J_1\cos(\vartheta_f-\vartheta_{b2}) - J_2\cos(\vartheta_f-\vartheta_{p1}) - J_3\cos(\vartheta_{p1}-\vartheta_{p2}) \end{aligned} \tag{9}$$

Here t_f, t_{p1}, t_{p2}, and t_{b2} are the thickness of the FL, P_1, P_2, and B_2 layers, respectively, K_{uf}, K_{up1}, K_{up2}, and K_{ub2} the anisotropy constant, ϑ_f, ϑ_{p1}, ϑ_{p2}, and ϑ_{b2} the angles of the magnetization. φ_b and φ_p are the pinning angles for B2 and P2 layers, respectively. All angles are defined with respect to the easy axis of free layer. M_{sb} and M_{sp} are the saturation magnetization of the B and P layers, respectively. H_p and H_b are the effective pinning fields of the P_2 and B_2 layers, respectively. J_1 is the interlayer coupling between FL and B_2 layers, and J_2 is the interlayer coupling between the FL and P_1 layers, and J_3 is the antiferromagnetic interlayer coupling between P_1 and P_2 layers.

We used the values $\lambda_0^{\uparrow} = 10.5$ nm, $\lambda_0^{\downarrow} = 0.5$ nm, $a^{\uparrow} = 0.0327$, $a^{\downarrow} = -0.00556$, $b^{\uparrow} = b^{\downarrow} = 0$. $M_s = 1543$ emu/cm^3 for CoFe, and $M_s = 800$ emu/cm^3 for NiFe. The parameters $K_{uf} = 10^3$ erg/cm^3, $K_{up1} = 2 \times 10^3$ erg/cm^3, $K_{up2} = 1.5 \times 10^4$ erg/cm^3, $K_{ub2} = 1.2 \times 10^4$ erg/cm^3, $H_{p2} = 240$ Oe, $H_{b2} = 300$ Oe, $J_1 = 8 \times 10^{-3}$ erg/cm^2, $J_2 = 10^{-3}$ erg/cm^2, and $J_3 = 0.12$ erg/cm^2 were chosen to response fit the experimental results.

For sample 3, $\varphi_p = 85°$ was obtained by curve fitting. The magnetization orientations of the FL as a function of applied field, $\theta_f(H)$, were obtained from the numerical calculations shown in Figs. 14(c), 14(f), and 14(i) for $\alpha = 80°, 90°$, and 100°, respectively. Figure 14(f) showed the smooth changes of θ_f with the applied field, which suggested that the magnetization of the FL rotated coherently with applied field. This type of FL reversal is labeled type-C reversal [27], corresponding to the coherent magnetization rotation, which was Barkhausen noise free. However, then a deviated from 90° a few degrees, as shown in Figs. 14(c) and 14(i), the $\theta_f(H)$ curves showed a discontinuous jump, suggesting that there existed a abrupt domain-wall motion during the magnetization reversal process. The magnetization jump gave rise to Barkhausen noise in a spin valve sensor despite the linear MR response of the sensor.

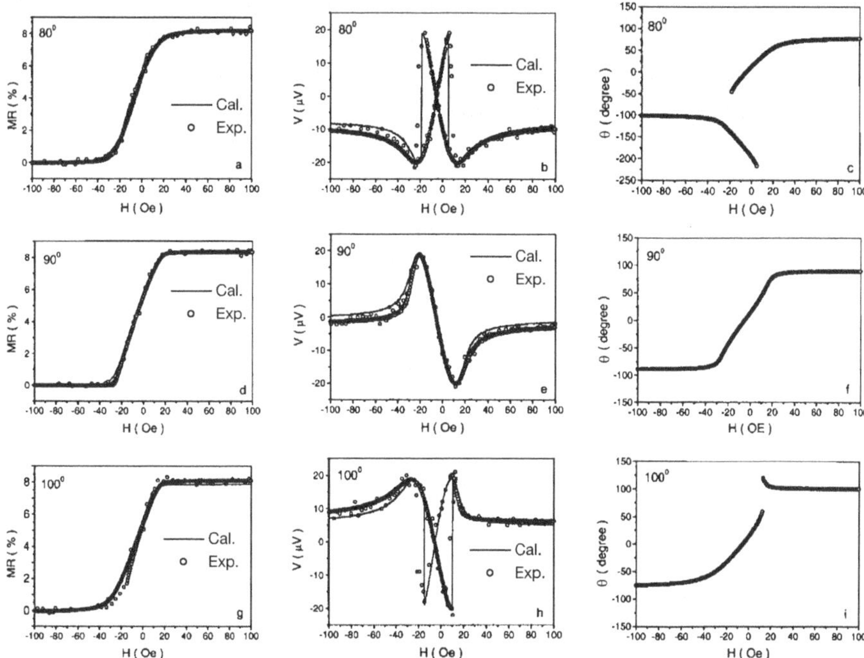

Fig. 14. Experimental (dotted) and calculated (solid line) MR and PHE responses of sample 3 in applied magnetic field at different α. First column shows the MR curves, the second column the PHE curves, and third column the corresponding simulated magnetization angles of the FL layer

For sample 4, $\varphi_p = 87°$ and $\varphi_b = -3°$ were obtained by curve fitting. The magnetization orientation of the FL as a function of applied field, represented by the $\theta_f(H)$ curve, was shown in Figs. 15(c), 15(f), and 15(i) for a of 80°, 90°, and 100°, respectively. Only the reversal mode of type C was observed for this sample for $70° < \alpha < 110°$. The FL appeared to remain in the single domain state in all three cases. It is evident that the interlayer coupling biasing field in the configuration kept the FL in a single domain state and constrained the magnetization reversal of the free layer in coherent rotation. The magnetization reversal of the FL was free from Barkhausen jumps. The major advantages of such a spin-engineered biasing scheme are that there is no "dead zone problem" as in PM biasing, the strength of biasing field can be controlled by varying the thickness of the Cu_1 spacer, the same AF material is used for biasing and pinning, and finally there is only one magnetic annealing to obtain the orthogonal biasing and pinning configuration of the FL and pinned layers.

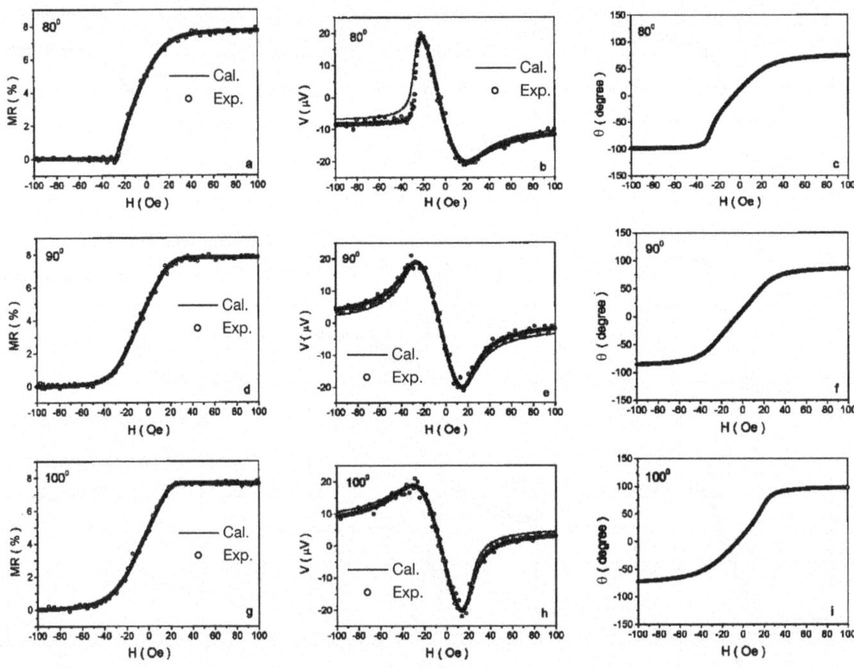

Fig. 15. Experimental (dotted) and calculated (solid line) MR and PHE responses of sample 4 in applied field at different a. First column shows the MR curves, the second column the PHE curves, and the third column the corresponding simulated magnetization angles of the FL layer

5 Conclusions

The basic principles, advantages and disadvantages of the existing domain biasing techniques were reviewed. The most commonly used domain bias techniques are permanent magnet biasing (PM biasing) and antiferromagnet exchange-tab biasing (AF exchange-tab biasing). However, these existing domain bias techniques are becoming obsolete as the continued miniaturization of sensors is approaching the nanometre regime. Several overlaid longitudinal domain bias schemes have been suggested to replace them. The longitudinal bias field can be obtained by the interlayer exchange coupling or long-range exchange coupling. However, there will face the problem that needs two AF materials with distinctly blocking temperature to achieve the orthogonal orientation of the longitudinal domain bias field and the transverse pinning field. A new spin-engineer domain bias scheme is developed from our recent work on interlayer exchange coupling and spin flop phenomena. The major advantages of this bias scheme are summarized as follows: a) One step magnetic annealing process to achieve orthogonal configuration for bias field and pinning field; b) No magnetic inactive zones in the sensor layer; c) Well defined trackwidth, no side reading problem; d) Bias field strength easy to control; e) The same AF material is used for biasing and pinning. This new domain bias scheme is promising for the nano-scale magneto-electronic devices.

References

1. C. H. Tsang, R. E. Fontana, Jr., T Lin, D. E. Heim, B. A. Gurney, and M. L. Williams, IBM J. Res. Dev. **42**. 103 (1998)
2. D.E. Heim, R. E. Fontana, Jr., C. Tsang, V. S. Speripsu, B. A Gurney, M.L. Williams, IEEE Trans Magn. **30**, 316 (1994)
3. C. Tsang, R. E. Fontana, T. Lin, D. E. Heim, V. S. Speriosu, B. A. Gurney,and M. L. Williams, IEEE Trans. Magn. **30**, 3801 (1994)
4. C Tsang, IEEE Trans. Magn. **25**, 3692 (1989)
5. Y. Shen, W. Chen, W. Jensen, D. Raipati, V. Retort, R. Rottmayer, S. Rudy, S. Tan and S. Yuan, IEEE Trans. Magn. **32**, 19 (1996)
6. A. J. Devashayam, J. Wang, and H. Hedge, J. Appl. Phys. **89**, 6615 (2000)
7. J. G. Zhu, Y. Zheng, S. Liao, IEEE Trans. Magn. **37**, 1723 (2001)
8. A. M. Mack, K. Subramanian, L. R. Pust, C. J. Rea, N. Amin, M.A. Seigler, S. Mao, S. Xue, and S. Gaangopadhyay, IEEE Trans. Magn. **37**, 1727 (2001)
9. T. Miyazaki and N. Tezuka, J. Magn. Magn. Mater. **139**, L231(1995)
10. J.S. Moodera, L. R. Kinder, T. M.Wong, and R. Meserey, Phys. Rev. Lett. **74**, 3237 (1995)
11. S. S. Parkin, K. P. Roche, M. G. Samant, P. M. Rice, R. B. Beyers, R. E. Scheuerlein, E. J. O'Sullivan, S. L. Brown, J. Buccigano, D. W. Abraham, Y. Lu, M. Rooks, P. L. Trouilloud, R. A. Wanner, and W. J. Gallagher, J. Appl. Phys. **85**, 5828 (1999)
12. Sining Mao, Yonghua Chen, Feng Liu, et al, IEEE Trans. Magn. **42**, 97 (2006)
13. M. Jullire, Phys. Lett. A **54**, 225 (1975)
14. W. H. Butler, X.-G. Zhang, T. C. Schulthess, and J. M. MacLaren, Phys. Rev. B **63**, 054416 (2001)
15. J. Mathon and A. Umersky, Phys. Rev. B **63**, 220403R (2001).
16. S. Yuasa, A. Fukushima, T. Nagahama, K. Ando, and Y. Suzuki, Jpn. J. Appl. Phys. **43**, L588 (2004).
17. S. Yuasa, T. Nagahama, A. Fukushima, Y. Suzuki, and K. Ando, Nat. Mater. **3**, 868 (2004)
18. S. S. P. Parkin, C. Kaiser, A. Panchula, P. M. Rice, B. Hughes, M. Samant, and S. H. Yang, Nat. Mater. **3**, 862 (2004)
19. D D. Djayaprawira, K Tsunekawa, M Nagai, H Maehara, S Yamagata, N Watanabe, S Yuasa, Y Suzuki, and K Ando, Appl. Phys. Lett. **86**, 092502 (2005)
20. K Tsunekawa, D D. Djayaprawira, M Nagai, H Maehara, S Yamagata, N Watanabe, S Yuasa, Y Suzuki, and K Ando, Appl. Phys. Lett. **87**, 072503 (2005)
21. E. Nakashio, J. Sugawara, S. Onoe, and S. Kumagai, J. ppl. Phys. **89**, 7356 (2001)
22. S. Mao, Z. gao, H. Xi, P. Kolbo, M. Plumer, L. Wang, A. Goyal, I. Jin, J. Chen, C. Hou, R. M. White, and E. Murdock., IEEE Trans. Magn. **38**, 26 (2002)

23. Z. Q. Lu and G. Pan, Appl. Phys. Lett. **80**, 3156 (2002)
24. J. G. Zhu and Y. Zheng, IEEE Trans. Magn. **34**, 1063 (1998)
25. H. C. Tong, C. Qian, L. Miloslavsky, S. Funada, X. Shi, F. Liu, and S. Dey, J. Appl. Phys. **98**, 5055 (2000)
26. Z. Q. Lu, G. Pan, A. Al-Jibouri, and Y. Zheng, J. Appl. Phys. **91**, 287 (2002)
27. Z. Q. Lu, G. Pan and A. Al-Jibouri, J. Appl Phys. **91**, 7116 (2002)
28. R.S. Beach, J. McCord, P. Webb,and D. Mauri, Appl. Phys. Lett. **80**, 4576 (2002)
29. Z. Q. Lu and G. Pan, Appl. Phys. Lett. **82**, 4107 (2003)
30. W. H. Meiklejohn and C. P. Bean, Phys. Rev. **105**, 904 (1957)
31. N. J. Goenmeijer, T. Ambrose, and C.L.Chien, J. Appl. Phys. **81**, 4999 (1997)
32. N. J. Goenmeijer, T. Ambrose, and C.L.Chien, Phys. Rev. Lett. **79**, 4270(1997)
33. B. Dieny, M. Li, S. H. Liao, C. Hong, and K, Ju, J. Appl. Phys. **87**, 3415 (2000)
34. J. Chen, S. Mao, J. Fernandez, T. Choey, ands. Hershfield, IEEE Trans. Magn. **26**, 2885 (2000)
35. T.G.S.M. Rijks, R. Coehoorn, M. J. M. de Jong, and W. J. M. de Jong, Phys. Rev. B **51**, 283 (1995)
36. J.P. King, J. N.Chapman, J. C. Kools, and M. F. Gillies, J. Phys. D **32**, 1087 (1999)
37. T. Lin, D. Mauri and P. M. Rice, J. M. M. M. **262**, 346 (2003)
38. M. L. Yan, Z. S. Shan, D. J. Sellmyer, and W. Y. Lai, J. Appl. Phys. **81**, 4785 (1997)

The manufacturer's authorised representative in the EU is Springer Nature Customer Service Centre GmbH, Europaplatz 3, 69115 Heidelberg, Germany. If you have any concerns regarding our products, please contact ProductSafety@springernature.com

Printed and bound by CPI Group (UK) Ltd, Croydon, CR0 4YY

15/07/2026

02167622-0002